Contents

Preface	vi
1. Selecting Wood to Burn	9
2. Firewood Cutting Tools	27
3. Wood for Your Woodpile	53
4. Seasoning Firewood	73
5. Wood Stoves, Fireplaces, and Other Wood Burners	95
6. Installing Wood Burning Systems	119
7. Operating Wood Stoves and Fireplaces	145
8. Wood Burning Appliance Maintenance and Safety	159
References	173
Metric Conversions	175
Index	176

This handbook provides basic information relating to the use of chain saws and to the installation and use of stoves, stovepipes, fireplaces, and chimneys. It is not intended to replace instructions or precautions given by the installer, maker, or vendor of any of these items. Chemicals and finishing materials referenced in this book likewise must be used in accordance with the manufacturer's instructions.

This handbook also presents information relating to the construction of stoves, stovepipes, fireplaces, and chimneys. This information is intended to merely acquaint the reader with these projects and it is not to be followed in actually designing or building these items. To insure quality and safety in design and construction, consult a professional engineer or builder.

While this handbook contains many safety tips, it is not possible to provide precautions for all possible hazards that might result. Standard and accepted safety procedures must be applied at all times.

Preface

If you are getting the notion of using wood to supplement or completely supply present home heating requirements, you are not alone. An increasing number of home owners throughout the nation are already burning wood in response to the rising cost of fossil fuels, such as coal, oil, and gas.

Looking back on the early days of this country, wood played a major role in providing fuel for heating and cooking in American households. In 1844, 95 percent of all energy needs in America were furnished by wood. By 1870, each person consumed about five tons, mainly for fuel. As the 20th Century approached, however, the use of wood for fuel began to decline. Although wood was still abundant, fossil fuels were becoming more available and were easier to handle. Exploitation and extensive use of these nonrenewable resources have resulted in projected shortages for the near future—thus generating the renewed interest in wood as a home heating fuel.

Man-made shortages as well as natural ones have brought back the burning of wood. Since the Arab oil embargo of 1973, people have been pushing toward the use of wood energy as an alternate home heating fuel. This movement has caused the resurrection or rebirth of an old American industry: the manufacture of wood burning heaters, stoves, furnaces, and a host of auxiliary products to serve the movement, such as wood splitters, heat exchangers, and prefabricated chimneys. Sales of chain saws have increased and the commercial fuelwood business has also reappeared largely as a result of this movement. Even the highest levels have been affected by the wood energy revival, with the installation of six wood stoves at the White House as well as several more at the presidential retreat in Camp David, Maryland.

Wood is formed by solar energy. Through the process of photosynthesis, sunlight converts carbon dioxide and water into organic material. When the tree dies, bacteria convert it back to carbon dioxide and water which are then returned to the soil. When a tree is burned, it releases these same ingredients: carbon dioxide and water, plus ash. If the burning process is thorough and efficient, the stored energy is released as heat and the nutrients are recycled back to the ground. Burning wood does not disrupt natural processes, it merely speeds them up.

Although all parts of a tree are combustible, certain pieces—the trunks, large limbs, and tops—are preferable for fuelwood because of variations in heating value, ease of handling, or other special problems. Roundwood accounts for the largest portion of fuelwood—more than 70 percent. It is either cut particularly for fuel or salvaged during clearing operations for rights of way, construction, farming, replacement of defective timber stands, timber stand improvement cuttings, or from logging residues. Dead or down timber also provides a considerable amount so that it is quite possible to obtain one's supply of fuelwood without having to cut a single living tree.

Unlike fossil fuels, trees are a renewable source of energy. Little or no significant environmental damages occur if the forest is properly maintained and only the amount of wood that can be replaced is taken. By promoting new growth in cutting areas, a wider diversity of wildlife habitats is created.

Wood has become one of the most popular alternative fuels. It can never totally replace coal, oil, or nuclear power, but it can complement these as well as ease the energy crunch in the future. Past use of wood has been a reflection of its low cost and ready availability rather than its overall advantages over other fuel types. This has been especially true in rural areas.

Comparisons of wood to other fuels, with their shortages and increasing costs, reveal that it does have certain advantages to many home owners. Some of these advantages are:

1. Wood is a relatively clean fuel.
2. Wood production is labor intensive; few cash costs are involved.
3. Wood is a safe fuel when used carefully.
4. Wood furnishes an independent heat source in the event of power failure or fuel shortage.
5. Wood is relatively simple to ignite and burns well when properly dried.
6. Wood, as previously mentioned, is a renewable and readily available resource; proper cutting procedures improve timber stand quality.

Burning wood also has disadvantages which have been largely responsible for the preference of other energy sources. Now that alternative fuels are scarcer and more expensive, some of these drawbacks are possibly less restrictive. Some of wood's disadvantages as a heating fuel are:

1. Wood is heavy and bulky in relation to its heat content, making it difficult to transport over long distances and awkward to store except in relatively large areas.
2. For best burning performance, wood should be dry. Time-consuming seasoning interrupts the production process as well as requires more long range planning than do fossil fuels.
3. Wood fires must be regularly tended and ashes must be removed. Stove pipes and chimneys need periodic cleanings to remove creosote and soot.
4. Building up a wood supply is not only strenuous work, it can be very dangerous unless done properly.
5. Wood possesses a relatively low heat value per unit of volume.
6. Burning speed is hard to control because of the irregularity in weight and heat values among various wood species.

In addition, aside from the time factor, obtaining one's own fuelwood makes sense provided: (1) one has the proper equipment (chain saw, axes, wedges, sawbuck, splitting maul, a pickup truck or trailer for hauling the wood); (2) the wood is located within a reasonable distance; and (3) the wood can be cut without any unnecessary difficulties. Often, the economics of acquiring one's own fuelwood may not be as obvious as first appears.

Fuel shortages could also affect the economics of wood cutting, but securing one's own wood supply may still prove cheaper than purchasing it, particularly if the home owner requires a substantial wood supply for the heating season. In the past, collecting firewood was simply a pleasurable outing for a family or group of friends and neighbors. To some degree, it still can be an enjoyable undertaking unless one is totally dependent on wood for heating. Then it becomes serious business.

This book was prepared as a guide to selecting, obtaining, cutting, splitting, and seasoning of firewood. In addition, there are chapters on the selection, installation, safe use, and maintenance of wood burners. In other words, FIREWOOD AND YOUR CHAIN SAW, as its name implies, is a complete book on the subject.

The compilation of this book required the help of many people. We wish to thank the U.S. Department of Energy (DOE), Pennsylvania State University (Cooperative Extension Service), American Forestry Association, National Park Service, U.S. Department of Agriculture (Forest Service), National Wildfire Coordinating Group, Northeast Regional Agricultural Engineering Service (NRAES), University of Vermont (Extension Service), National Fire Protection Association, as well as various stove and fireplace manufacturers, for their permission to use certain written technical information of theirs that appears in this book. We would also like to thank Carl Markle of Robert Scharff and Associates, Ltd. for the many hours he spent in researching and coordinating the material that has made this book possible. Thanks again to everyone who helped in welcoming you to the wonderful world of wood heating.

Selecting Wood to Burn 1

Choosing a kind of firewood to burn in your wood burner is much like selecting a favorite wine or cheese, since each wood species can offer something different in aroma or heat value. The fuelwood connoisseurs will want to choose their wood carefully and weigh their needs and tastes before building their fire. But, before considering the characteristics of fuelwoods, let us first look at what wood is all about.

ANATOMY OF A TREE

The tree is an extremely complex organism composed of living and dead cells (Fig. 1-1). The size and arrangement of the cells determine the grain of the wood and many of its properties. Examine a freshly cut tree stump—you will see that the millions of small cells are arranged in circular rings around the pith or center of the tree (Fig. 1-2). These rings are caused by a difference in the rate of growth of the tree during the various seasons of the year. In spring, a tree grows rapidly and builds up a thick layer of comparatively soft, large cells which appear in the cross section of the trunk as the light colored annual rings.

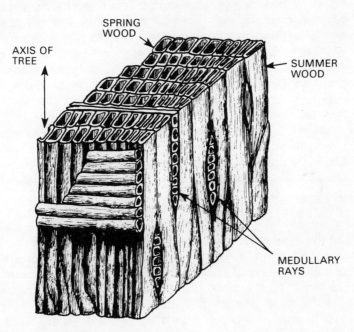

Fig. 1-1: Structure of wood.

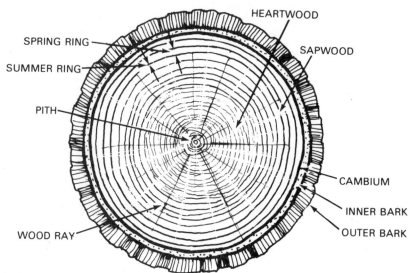

Fig. 1-2: Cross section of a tree.

As the weather becomes warmer during the early summer, the rate of growth slows and the summer growth forms cells that are more closely packed. These pairs of concentric springwood and summerwood rows form the annual rings which can be counted to find out the age of the tree. Because of climatic conditions, some trees, such as oak and walnut, have more distinctive rings than maple and birch. White pine is so uniform that you can hardly distinguish the rings, while many other softwoods have a very pronounced contrast between summerwood and springwood; this makes it easy to distinguish the rings.

The sapwood of a tree is the outer section of the tree between the heartwood (darker center wood) and the bark. The sapwood is lighter in color than the heartwood, but, as it gradually changes to heartwood on the inside and as new layers are formed, it becomes darker. Depending upon the type of tree, it requires from nine to 36 years to transform sapwood into heartwood.

The cambium layer, the boundary between the sapwood and the bark, is the thin layer where new sapwood cells form. Medullary rays are radial lines of wood cells consisting of threads of pith which serve as lines of communication between the central cylinder of the tree and the cambium layer. They are especially prominent in oak.

When a tree is sawed lengthwise, the annual ring forms a pattern, called the grain of the wood. A number of terms are used to describe the various grain conditions. If the cells which form the grain are closely packed and small, the wood is said to be fine grained or close grained. Maple and birch are excellent examples of this type of wood. If the cells are large, open, and porous, the wood is coarse grained or open grained, as in oak and walnut. When the wood cells and fibers are comparatively straight and parallel to the trunk of the tree, the wood is said to be straight grained. If the grain is crooked, slanting, or twisted, it is said to be cross grained. It is the arrangement, direction, size, and color of the wood cells that give the grain of each wood its characteristic appearance.

Live or green wood can contain up to 65 percent moisture. To be of value as firewood, it must be well seasoned as explained later in the book. Briefly, however, seasoning of wood is accomplished by removing the moisture from the millions of small and large cells of which wood is composed. Moisture (water or sap) occurs in two separate forms: free water and embedded water. Free water is the amount of moisture the individual cells contain. Embedded water is the moisture absorbed by the cell walls. During drying or seasoning, the free water in the individual cells evaporates until a minimum amount of moisture is left. The point at which this minimum moisture remains is called the fiber saturation point. The moisture content of this point varies from 20 to 30 percent. Below the fiber saturation point, the embedded water is extracted from the porous cell walls; this process causes a reduction of the thickness of the walls. Wood shrinks across the grain when the moisture content is lowered below the fiber saturation point. Shrinking and swelling of the wood cells caused by varying amounts of moisture change the size of the cells.

Although scientific nomenclature defines two types of woods—hard and soft—fuelwood experts add a third classification—fruitwood. The terms "hardwood" and "softwood" have no reference to the softness or hardness of the wood. For instance, Douglas fir, a softwood, is much harder than poplar, which is a hardwood.

Softwood comes from cone-bearing (coniferous) trees, those with scale-like or needle-like leaves, such as fir, pine, cedar, redwood, and spruce. Hardwood comes from non-cone-bearing (deciduous) trees, those that have broad leaves, such as oak, maple, birch, and elm. Fruitwoods—apple, cherry, and citrus—are really hardwoods that have special fuelwood qualities.

HEATING CHARACTERISTICS OF WOOD

The most important fuelwood characteristic that we should be concerned with is the heating value. Actually, the heat that is derived from the combustion of wood depends upon the concentration of: (1) woody material; (2) resin; (3) ash; and (4) water. The first three features vary depending on the tree species and its growth rate, while the latter depends on the species, the season in which the tree was cut, and the seasoning procedures used. In general, the heaviest woods, when seasoned, have the greatest heating value. Lighter woods give about the same heat value per pound as heavier hardwoods, but, because they are less dense, they give less heat per cord or cubic foot. For example, black locust wood has almost twice the density and weight as the same volume of Eastern white pine, and it has twice as much heat energy. Of course, excessive moisture in firewood reduces its fuel value. Wood containing 20 percent moisture content—which is considered the normal for air dried condition—will lose about 3.5 percent of the net heat content of dry wood as a result of moisture evaporation losses. In contrast, wood containing 50 percent moisture will lose about 8.6 percent of the net heat due to the presence of moisture. The heating values per air-dried standard cord of numerous varieties of wood as compared to other fuels are given in Table 1-1.

A frequently asked question is: How does wood compare with other fuels as a heat source? By using the values contained in Table 1-1, it is possible to compare the cost of various fuels. For example, if No. 2 fuel oil is selling for $1.01 per gallon, what would be the equivalent cost per cord for air-dried red oak?

Table 1-1:
HEAT AVAILABLE FROM WOOD COMPARED WITH EQUIVALENT AMOUNT OF OTHER FUELS (see page 14 for key)

Wood (Standard Cord) a	Type of Wood	Pounds per cubic feet b	Total Btu's/cord (millions) c	Available Btu's/cord (millions) d	Anthracite Pea Coal (tons) e	# 2 Fuel Oil (gallons) f	Natural Gas (100 cubic ft.) g	LP Gas (gallons) h	Electricity (kilowatt hours) i
Alder, red	Hard	25.5	15.65	7.80	0.72	89	107	112	2355
Apple	Fruit	44.0	27.20	13.60	1.21	148	179	195	3935
Ash, black	Hard	30.5	18.70	9.35	0.92	113	137	150	3030
Ash, green	Hard	35.0	21.50	10.75	1.03	126	153	167	3370
Ash, Oregon	Hard	34.5	21.20	10.60	1.02	124	151	165	3330
Ash, white	Hard	37.5	23.00	12.50	1.09	141	161	176	3570
Aspen, bigtooth	Hard	24.0	14.75	7.40	0.67	82	100	109	2200
Aspen, quaking	Hard	23.5	14.45	7.20	0.65	79	96	104	2060
Basswood, American	Hard	23.0	14.10	7.05	0.62	76	92	100	2020
Beech, American	Hard	40.0	24.60	12.30	1.11	136	165	180	3635
Birch, black	Hard	40.5	24.80	12.40	1.17	144	175	190	3840
Birch, white (paper)	Hard	34.5	21.20	10.60	1.02	126	156	168	3390
Birch, yellow	Hard	38.5	23.65	11.80	1.12	138	168	183	3390
Butternut	Hard	23.5	14.40	7.20	0.70	86	104	113	2285
Cedar, red, Eastern	Soft	29.0	17.80	8.90	0.87	107	130	141	2870
Cedar, red, Western	Soft	20.0	12.30	6.15	0.64	80	98	108	2195
Cedar, white (arbor vitae)	Soft	19.0	11.70	5.90	0.62	77	94	104	2120
Cherry, black	Fruit	31.0	19.00	9.50	0.92	113	137	149	3020
Citrus	Fruit	41.0	25.40	12.70	1.06	131	160	174	3530
Cottonwood, black (Western)	Hard	22.0	13.50	6.75	0.61	75	91	99	2015
Cottonwood, Eastern	Hard	25.0	15.40	7.70	0.71	87	105	114	2315
Cypress, bald	Soft	28.5	17.50	8.75	0.80	99	120	130	2640
Dogwood, flowering	Hard	46.0	27.40	13.70	1.19	161	178	193	3905
Elm, American	Hard	31.0	19.00	9.50	0.89	110	133	145	2930
Elm, Rock	Hard	39.0	23.95	12.00	1.09	134	161	175	3530
Eucalyptus	Hard	28.0	17.20	8.60	0.78	97	118	128	2600
Fir, balsam	Soft	22.5	13.80	6.90	0.65	81	98	106	2130
Fir, Douglas	Soft	30.0	18.40	9.20	0.89	110	133	145	2945
Fir, noble	Soft	24.0	14.70	7.35	0.69	84	101	110	2240
Gum, black (tupelo)	Hard	31.0	19.00	9.50	0.92	113	137	149	3020

Selecting Wood to Burn   13

Wood (Standard Cord) a	Type of Wood	Pounds per cubic feet b	Total Btu's/cord (millions) c	Available Btu's/cord (millions) d	Anthracite Pea Coal (tons) e	# 2 Fuel Oil (gallons) f	Natural Gas (100 cubic ft.) g	LP Gas (gallons) h	Electricity (kilowatt hours) i
Gum, sweet (red)	Hard	32.5	20.00	10.00	0.93	114	138	151	3040
Hackberry	Fruit	33.0	20.30	10.15	0.97	119	144	156	3170
Hemlock, Eastern	Soft	25.0	15.40	7.70	0.71	87	105	114	2315
Hemlock, Western	Soft	28.0	17.20	8.60	0.78	97	118	128	2600
Hickory, bitternut	Hard	41.0	25.20	12.60	1.06	131	160	174	3530
Hickory, shagbark	Hard	49.0	30.60	15.30	1.26	155	188	204	4130
Ironwood, hardhack (hophornbeam)	Hard	48.5	30.10	15.05	1.25	154	187	203	4105
Juniper	Soft	24.5	15.20	7.60	0.70	87	105	114	2280
Locust, black	Hard	44.0	26.50	13.25	1.28	158	190	206	4160
Locust, honey	Hard	37.5	23.00	12.50	1.14	141	171	186	3750
Magnolia, Southern	Hard	31.0	19.00	9.50	0.88	109	133	144	2940
Maple, red	Hard	33.5	20.60	10.30	0.98	116	142	153	3115
Maple, silver	Hard	29.0	17.90	8.95	0.83	103	126	137	2790
Maple, sugar	Hard	38.5	23.65	11.80	1.06	131	159	172	3490
Oak, Arizona White (emory)	Hard	40.0	24.60	12.30	1.13	138	168	182	3680
Oak, live	Hard	55.0	33.80	16.95	1.50	183	220	238	4800
Oak, pin	Hard	39.0	24.00	12.00	1.10	135	164	178	3605
Oak, red	Hard	39.0	24.10	12.05	1.10	135	164	178	3605
Oak, white	Hard	42.5	26.10	13.05	1.18	145	176	191	3870
Osage Orange	Fruit	44.0	27.20	13.60	1.23	150	182	197	3980
Peach	Fruit	41.5	25.90	12.95	1.07	133	162	176	3570
Pear	Fruit	40.5	25.20	12.60	1.04	130	158	172	3495
Persimmon	Fruit	47.0	28.70	14.35	1.31	159	193	208	4205
Pine, jack (gray)	Soft	27.0	16.60	8.30	0.76	93	116	125	2465
Pine, loblolly	Soft	32.0	19.65	9.85	0.91	117	136	149	3005
Pine, lodgepole	Soft	25.5	15.65	7.85	0.72	89	107	116	2355
Pine, longleaf	Soft	37.0	22.75	11.40	1.04	128	154	168	3380
Pine, pinyon (nut)	Soft	39.0	24.20	12.10	1.09	134	161	175	3530
Pine, pitch	Soft	32.5	20.00	10.00	0.93	114	138	151	3040
Pine, pond	Soft	35.0	21.50	10.75	0.99	126	147	160	3230
Pine, Ponderosa	Soft	25.0	15.50	7.75	0.71	87	105	114	2315
Pine, red (Norway)	Soft	28.5	17.50	8.75	0.80	99	120	130	2640
Pine, shortleaf	Soft	32.0	19.65	9.85	0.91	117	136	149	3005

Wood (Standard Cord) a	Type of Wood	Pounds per cubic feet b	Total Btu's/cord (millions) c	Available Btu's/cord (millions) d	Anthracite Pea Coal (tons) e	# 2 Fuel Oil (gallons) f	Natural Gas (100 cubic ft.) g	LP Gas (gallons) h	Electricity (kilowatt hours) i
Pine, slash	Soft	37.0	22.75	11.40	1.04	128	154	168	3380
Pine, sugar	Soft	22.5	13.80	6.90	0.66	82	99	108	2180
Pine, white/Eastern	Soft	22.0	13.25	6.65	0.65	80	97	106	2140
Pine, white/Western	Soft	23.5	14.45	7.30	0.69	85	103	112	2255
Pine, whitebark	Soft	25.0	15.55	7.75	0.71	87	105	114	2315
Plum	Fruit	40.0	25.00	12.50	1.13	138	168	182	3680
Poplar, balsam	Hard	21.0	12.90	6.45	0.62	77	93	102	2100
Redwood	Soft	22.0	13.50	6.75	0.65	80	97	106	2140
Sassafras	Hard	28.5	17.50	8.75	0.80	99	120	130	2640
Shadbush, Eastern	Hard	44.0	26.80	13.40	1.23	150	182	197	3980
Spruce, Englemann	Soft	22.0	13.50	6.75	0.65	80	97	106	2140
Spruce, red	Soft	25.5	15.65	7.80	0.72	89	107	112	2355
Spruce, Sitka	Soft	25.0	15.40	7.70	0.71	87	105	114	2315
Spruce, white	Soft	25.0	15.40	7.70	0.71	87	105	114	2315
Sycamore, American	Hard	30.5	18.70	9.35	0.88	108	131	144	2890
Tamarack (American larch)	Soft	33.0	20.30	10.15	0.97	119	145	158	3195
Tamarack (Western larch)	Soft	32.5	20.00	10.00	0.93	114	138	151	3040
Walnut, black	Hard	34.5	21.20	10.60	1.01	124	151	164	3310
Willow, black	Hard	24.0	14.70	7.35	0.69	84	101	110	2240
Willow, brittle	Hard	23.0	14.00	7.00	0.67	81	98	106	2165
Yew	Soft	40.0	24.00	12.30	1.13	138	168	182	3680

a—Standard cord of wood: 128 cubic feet total space. Assume average cord of wood contains 80 cubic feet of solid wood.
b—Weight per cubic foot calculated at 20% moisture content, air-dried.
c—Total heat content at 20% moisture is 6900 Btu/pound.
d—Available heat assumes wood burner is operating at 50% efficiency.
e—Anthracite coal contains 28,000,000 Btu/ton, but available heat is only 22,000,000 Btu/ton, or 11,000 Btu/pound. Assume 60% efficiency of furnace.
f—One gallon #2 fuel oil contains 140,000 Btu. At 65% efficiency of the oil burner, heat energy is 91,000 Btu.
g—Natural gas contains 100,000 Btu/100 cubic feet. With furnace at 75% efficiency, heat energy is 75,000 Btu.
h—Liquid propane contains 92,000 Btu/gallon, but when burned in appliance at 75% efficiency, heat energy is 69,000 Btu.
i—Total heat is 3412 Btu/kilowatt hour. Efficiency is 100%.

Heat from 135 gallons of No. 2 fuel oil is equivalent to the available heat from one cord of red oak. Therefore, the homeowner could afford to pay $136.35 for a standard cord of red oak firewood. Of course, there is considerably more work associated with burning wood. If the labor cost is to be considered, it would have to be added to the wood cost when comparing fuel costs. If one is cutting his own firewood, then costs of chain saw, fuel, transportation, and miscellaneous tools should also be considered. A good rule of thumb for determining the comparison between wood and oil is: Hardwood fuel should be no higher (in dollars per cord) than one and one-half times the price of oil (in cents per gallon).

For an all-electric home, comparing costs between electricity and wood is rather interesting. For example, suppose electricity costs $.04 per kilowatt hour, one could afford to pay $145.36 per standard cord for red oak fuelwood, based on 3,605 kilowatt hours of electricity being equivalent to a cord of red oak.

Another frequently asked question concerns the amount of wood needed to heat a home for the year. Suppose you burned 1,400 gallons of No. 2 fuel oil last year. Using this as a guide, we can estimate the wood required by using values from Table 1-1. Number 2 fuel oil has 91,000 Btu's per gallon when burned at 65 percent efficiency. This amount of fuel oil will yield 127,400,000 Btu's of heat energy. An air-dried cord of red oak burned at 50 percent efficiency will yield 12,050,000 Btu's. Therefore, the equivalent amount of red oak fuelwood needed to heat the home is 10.3 standard cords.

Moisture Content of Wood

As mentioned earlier, green wood is not only more difficult to burn, but it is wasteful of energy. Any moisture in the wood reduces the recoverable heat because water absorbs heat in the process of being changed to steam. The net heat from a pound of completely dry (no moisture) hardwood is about 7,950 Btu. All wood has some moisture in it which reduces the net usable heat at a rate of 1,200 Btu per pound of water. Also, the resin in green wood causes a dangerous buildup of creosote inside the wood burner's stovepipe and/or chimney. (Creosote is the black, foul smelling liquid which contains wood tars, acids, alcohols, and water. A product of incomplete wood combustion, creosote generally occurs as a buildup on the inside of stovepipes and chimney flues. If serious enough, it may completely block the flow of chimney gases or provide fuel for serious chimney fires. Creosote formation can be minimized by periodically burning hot fires and using wood that is thoroughly air dried. More on the problems of creosote and how to solve them are given in Chapter 5.)

The moisture in the wood of living trees varies among species, within a species, and even within a single individual. Frequently, there is a significant difference between the quantity of moisture contained in the central column of heartwood of a tree and the outer layers of sapwood which are surrounded with bark. For example, freshly cut American beech has been found to have a heartwood moisture content of 72 percent. In contrast, heartwood moisture contents in American elm, northern red oak, and white ash are 95, 80, and 46 percent respectively. If you cut trees in summer, let them lie for a week. The leaves will draw moisture from the wood and dry it more quickly than if you limb the tree immediately.

Some woods, such as ash, beech, and Douglas fir, burn rather well after being only moderately seasoned, but to those who desire the maximum heating value available, proper seasoning and storage is a must. Seasoning time for green wood depends on a number of factors, but in most areas at least six, and preferably 12, months are necessary. With certain seasoning set-ups, such as solar driers, and in certain drier climates of the southwest and west, three months is often sufficient.

Methods of determining whether or not wood is dry and seasoning fuelwood are fully described in Chapter 4.

OTHER CHARACTERISTICS OF WOOD

When considering the type of wood to burn, other characteristics in addition to heat value are often important (see Table 1-2). These characteristics include: (1) ease of splitting; (2) ease of ignition and burning; (3) coaling qualities (ability to form long-lived coals); (4) extent of sparking; (5) extent of smoking; (6) aroma qualities; and (7) resistance to decay.

If you cut your own firewood or split kindling and fuel logs, the splitting characteristics of wood are very important. Short lengths of straight-grained, knot-free wood will split easily. Green wood and softwoods usually, but not always, split more easily than dry wood and hardwoods. Sometimes, frozen wood splits easily. Straight-grained cottonwood, aspen, fir, and pine split easily and are best for kindling. In contrast, woods with interlocking grain, like American elm and sycamore, may be very difficult to split and should be avoided if possible. (This, of course, applies only when the size of the firewood necessitates splitting.)

Softwoods, being resinous, are easy to ignite and burn rapidly with a high, hot flame. However, they burn out quickly and require frequent attention. Hardwoods, on the other hand, are generally more difficult to ignite, burn less vigorously and with a shorter flame, but last longer and produce more coals than softwoods. White birch typically is easy to ignite due to its papery, resinous bark. Oak gives the most uniform and shortest flame and produces steady, glowing coals. Generally, heavy-weight hardwoods and fruitwoods, because of their slower burning characteristics, provide the best coaling qualities.

Some resinous softwoods, such as cedar, juniper, larch, hemlock, and spruce, contain moisture pockets which can be troublesome. Upon heating, trapped gases and water vapor build up pressure in these pockets, resulting in "pops" which throw sparks. Such sparking can be a potential fire hazard especially in fireplaces without proper screens. Since green woods spark with greater intensity than dry woods, it is wise to reduce the moisture content of wood as much as possible before burning.

All woods smoke to some degree. As a rule, resinous softwoods produce more smoke than hardwoods or fruitwoods. A properly designed and operated wood burner can usually burn woods that produce heavy smoke. Most woods produce little or no aroma. Also, people react differently to different wood species; some may detect aroma while others do not. Only fruitwood and pines produce fragrances that most people can detect.

When the moisture in wood is excessive (25 percent or more), the wood can be susceptible to decay. Decay is caused by fungi, small plant organisms which feed on wood until it becomes soft and punky—not suitable for burning. That is, decayed or rotten wood is very low in heat value and makes a very poor fuel.

Selecting Wood to Burn 17

Table 1-2: QUALITY CHARACTERISTICS OF COMMONLY BURNED WOODS (see page 20 for key)

Name of Wood	Ease of Splitting	Ease of Ignition and Burning	Coaling Qualities	Extent of Sparking	Extent of Smoking	Aroma Qualities	Resistance to Decay
Alder, red	E	F	F	L	L	S	P
Apple	H	P	G	L	L	E	G
Ash, black	E	F	G	L	L	S	P
Ash, green	E	F	G	L	L	S	P
Ash, Oregon	E	P	G	L	L	S	P
Ash, white	E	P	G	L	L	S	P
Aspen, bigtooth	E	G	F	L	L	S	P
Aspen, quaking	E	G	F	L	L	S	P
Basswood	E	G	P	L	M	S	P
Beech, American	H	P	G	L	M	S	P
Birch, black	E	P	G	M	L	S	P
Birch, white (paper)	H	E	G	M	M	S	P
Birch, yellow	H	G	G	M	L	S	P
Butternut	E	G	G	G	M	G	G
Cedar, red (Eastern)	E	E	P	M	M	G	G
Cedar, red (Western)	E	G	P	G	M	S	G
Cedar, white (arbor vitae)	E	E	P	L	L	G	G
Cherry, black	E	P	E	L	L	E	G
Citrus	E	P	E	L	M	E	P
Cottonwood, black (Western)	E	G	F	L	M	S	P
Cottonwood, Eastern	E	G	F	L	M	S	P
Cypress, bald	E	F	F	L	L	F	F
Dogwood, flowering	E	E	E	L	L	F	F
Elm, American	VH	F	E	L	L	S	P
Elm, Rock	VH	P	E	L	L	S	F
Eucalyptus	H	P	F	G	M	E	P

Firewood and Your Chain Saw

Name of Wood	Ease of Splitting	Ease of Ignition and Burning	Coaling Qualities	Extent of Sparking	Extent of Smoking	Aroma Qualities	Resistance to Decay
Fir, balsam	E	G	P	L	M	S	P
Fir, Douglas	E	E	F	L	G	S	F
Fir, noble	E	E	F	M	M	S	P
Gum, black (Tupelo)	H	P	E	L	L	S	P
Gum, sweet (red)	H	P	G	L	L	F	P
Hackberry	H	G	G	L	L	G	P
Hemlock, Eastern	H	G	P	L	M	G	P
Hemlock, Western	H	P	E	L	M	S	P
Hickory, bitternut	H	P	E	L	L	S	P
Hickory, shagbark	H	P	E	L	L	S	P
Ironwood, hardhack (hophornbeam)	VH		E			S	G
Juniper	E	P	E	L	M	G	G
Locust, black	H	E	G	M	L	S	E
Locust, honey	H	P	E	L	L	F	F
Magnolia, Southern	E	G	G	L	L	G	P
Maple, red	H	F	E	L	L	G	P
Maple, silver	H	F	G	L	L	F	P
Maple, sugar	F	P	E	L	L	S	G
Oak, Arizona white (emory)	H	F	E	L	L	F	G
Oak, live	H	F	E	L	L	F	P
Oak, pin	H	P	E	L	L	F	G
Oak, red	H	P	E	L	L	F	P
Oak, white	H	P	E	L	L	F	G
Osage orange	VH	F	E	L	L	F	E
Peach	H	G	F	L	L	G	E
Pear	H	G	F	L	L	E	G
Persimmon	H	P	E	L	L	G	G

Selecting Wood to Burn 19

Name of Wood	Ease of Splitting	Ease of Ignition and Burning	Coaling Qualities	Extent of Sparking	Extent of Smoking	Aroma Qualities	Resistance to Decay
Pine, jack (gray, dwarf)	E	E	P	G	M	E	P
Pine, loblolly	E	E	F	G	G	E	P
Pine, lodgepole	E	E	F	G	G	E	P
Pine, longleaf	E	E	F	G	G	E	F
Pine, pinyon (nut)	E	E	F	G	M	E	P
Pine, pitch	E	E	F	G	G	E	P
Pine, pond	E	G	P	G	G	E	P
Pine, ponderosa	E	E	P	G	M	E	P
Pine, red (Norway)	E	E	P	G	G	E	P
Pine, shortleaf	E	E	P	G	G	E	F
Pine, slash	E	E	F	G	M	E	P
Pine, sugar	E	E	P	G	M	E	F
Pine, white/Eastern	E	E	P	G	M	E	P
Pine, white/Western	E	E	P	G	M	E	P
Pine, whitebark	H	F	F	G	L	E	G
Plum	E	G	P	M	M	S	P
Poplar, balsam	E	G	P	L	M	S	G
Redwood	E	P	P	L	L	E	G
Sassafras	F	F	P	L	L	F	P
Shadbush, Eastern	F	G	G	G	M	G	P
Spruce, Englemann	E	G	P	G	M	G	P
Spruce, red	E	G	P	G	M	G	P
Spruce, Sitka	E	G	P	G	M	S	P
Sycamore, American	VH	F	G	L	M	G	P
Tamarack (Eastern larch)	E	E	F	M	M	G	F
Tamarack (Western larch)	F	P	F	M	M	G	G
Walnut, black	F	P	G	L	L	G	G
Willow, black	E	G	P	L	L	S	P

Name of Wood	Ease of Splitting	Ease of Ignition and Burning	Coaling Qualities	Extent of Sparking	Extent of Smoking	Aroma Qualities	Resistance to Decay
Willow, brittle	E	G	P	L	L	S	P
Yew	H	F	G	L	L	G	E

1. All wood types are satisfactory as fuel; no common wood is truly unsuitable as fuel. The best practice is to burn whatever is readily available.
2. While many woods are listed as fuelwood for sake of information, in practice such woods as dogwood, eucalyptus, magnolia, osage orange, persimmon, redwood, and yew are seldom burned. Burn only the limbs of the black walnut; sawmills pay good money for the trunks.
3. Splitting Key: E = easy; F = fair; H = hard; VH = very hard. In fact, there is no "easy" splitting if it is done without the aid of power equipment. In practice, some woods split much easier than others.
4. Starting (igniting) Key: E = excellent; G = good; F = fair; P = poor. In general, the harder (denser) the wood, the harder it is to start. Softwoods make for good starting wood because of the high resin content.
5. Coaling Key: E = excellent; G = good; F = fair; P = poor. In general, the denser the wood, the better and longer-lasting are the coals. Fruitwoods make excellent coals.
6. Sparking Key: G = great; M = moderate; L = little. Green woods will spark with greater intensity than dry woods. The pines and cedars are noted for their sparking.
7. Smoking Key: G = great; M = moderate; L = little. As with sparking, green woods produce much more smoke than dry woods. The softwoods usually produce more smoke than hardwoods.
8. Aroma Key: E = excellent; G = good; F = fair; S = slight. Most woods produce little or no aroma. Some people react differently to different woods and may detect aroma where others do not. Only the pines and fruitwoods can be recommended for aroma.
9. Decay Key: E = excellent; G = good; F = fair; P = poor. The sapwood of all species is subject to decay. Decay may be greatly retarded by drying the wood, allowing good air circulation, and keeping the wood off the ground. Burn the most susceptible woods first, saving the resistant types for later burning.
10. While the live oak is listed as a hardwood, in fact it is an evergreen.
11. The Douglas fir is listed under the "firs" for convenience. In fact, it is not a true fir.

Selecting Wood to Burn 21

Chances of decay, of course, can be greatly reduced by drying the wood properly, allowing good air circulation in the wood pile, and keeping the wood off the ground (see Chapter 4 for details). Since woods vary in their resistance to decay, burn the most susceptible species first, saving the more resistant types for later.

Tree identification can sometimes be difficult, especially after the leaves have fallen. Some good fuel trees, such as the ashes, beeches, oaks, and hickories, are easily identified by their distinctive barks. Examining the bark of other trees is not of much value. The weathering effects of rain, sun, cold, heat, and wind tend to color the bark of the maple, poplar, gray birch, and spruce in much the same manner, and they all appear spotted, mottled, and covered with lichen and algae.

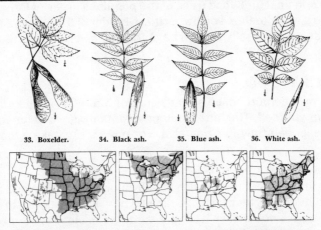

33. Boxelder.   34. Black ash.   35. Blue ash.   36. White ash.

33. Boxelder, *Acer negundo* L. (California boxelder, ashleaf maple, Manitoba maple, boxelder maple). Medium sized tree, including its varieties widely distributed across the United States and adjacent Canada. Bark gray or brown, thin, with narrow ridges and fissures. Twigs green. Leaves paired, compound, with usually 3 or 5, rarely 7 or 9, oval or lance-oblong leaflets 2–4 in (5–10 cm) long, long-pointed, coarsely and sharply toothed, bright green, nearly smooth or hairy. Key fruits 1–1½ in (2.5–4 cm) long, paired and in clusters, maturing in fall.
Principal uses: Shelterbelts. Shade tree.

    VV. Leaflets 5–11, bluntly toothed or without teeth, with veins curved within edges (fruits clustered but not in pairs, long-winged "keys")—**Ash** (*Fraxinus*).
    W. Leaflets without stalks.

34. Black ash, *Fraxinus nigra* Marsh. (swamp ash, water ash, brown ash, hoop ash, basket ash). Medium-sized to large tree of wet soils in northeastern quarter of United States and adjacent Canada. Bark gray, scaly or fissured. Leaves paired, compound, 12–16 in (30–40 cm) long, with 7–11 stalkless, oblong or broadly lance-shaped leaflets 3–5 in (7.5–13 cm) long, long-pointed, finely toothed, with tufted hairs beneath. Key fruits 1–1½ in (2.5–4 cm) long, ⅜ in (1 cm) wide, flat, with wing extending to base.
Principal uses: Same as No. 36.

    WW. Leaflets with stalks.

35. Blue ash, *Fraxinus quadrangulata* Michx. Medium-sized to large tree of Central States, chiefly Ohio and Mississippi Valley regions; also in southern Ontario. Bark gray, fissured, with scaly and shaggy plates. Twigs 4-angled and more or less winged. Leaves paired, compound, 8–12 in (20–30 cm) long, with 7–11 short-stalked, oval or lance-shaped leaflets 2½–5 in (6–13 cm) long, long-pointed, toothed. Key fruits 1¼–2 in (3–5 cm) long, ⅜–½ in (1–1.2 cm) wide, oblong, with wing extending to base.
Principal uses: Same as No. 36.

36. White ash, *Fraxinus americana* L. (American ash, Biltmore ash; *F. biltmoreana* Beadle). Large tree of eastern half of United States and adjacent Canada. Bark gray, with deep, diamond-shaped fissures and narrow, forking ridges. Leaves paired, compound, 8–12 in (20–30 cm) long, with 5–9, usually 7, stalked, oval or broadly lance-shaped leaflets 2½–5 in (6–13 cm) long, long- or short-pointed, edges sometimes slightly toothed, hairless or hairy beneath. Key fruits 1–2 in (2.5–5 cm) long and ¼ in (6 mm) wide, with wing at end.
Principal uses: Furniture, flooring, millwork, paneling, sporting and athletic goods, hand-tools, and boxes and crates. Shade tree.

Fig. 1-3: Typical page from the Department of Agriculture's Handbook No. 519 mentioned on page 22.

One of the easiest ways to become an expert in tree identification is to obtain a good book on the subject (there are several on the market) from your local bookstore. Or, you can obtain a copy of the U.S. Department of Agriculture's Handbook No. 519—*Important Forest Trees of the United States*—from the Superintendent of Documents, U.S. Government Printing Office, Washington, D.C. 20402. We have reproduced a page from this Forest Service publication (Fig. 1-3) to point out the various items that help to identify trees.

Species of trees vary throughout the United States. As shown in Fig. 1-4, the U.S. Forest Service has divided the country into various administrative regions. The dominant trees (Table 1-3) are those varieties likely to be found throughout the given region. Other species may be plentiful in certain localities. For instance, the principal fuelwood trees of the populous areas of Canada's southern border are similar to those of the northern border of the United States. Other sections of Canada generally tend to have the same varieties of trees as those found in Alaska.

OBTAINING WOOD

Fuelwoods may be purchased already cut or gotten from sources where you must cut them yourself. The latter method of obtaining wood for your stove or fireplace is described fully in Chapter 3. For now, let us take a look at buying firewood.

Buying Wood

While cutting your own firewood is usually more rewarding (and a great deal cheaper) than buying it, there may be times when it is necessary to do just this. But, buying wood is not always a simple matter. You can buy your wood from a dealer that you trust implicitly, or you must become an educated consumer, either by choice or by trial and error. Let us assume that you would prefer not to "get burned" before the wood gets into your stove and begin your education here. Five choices present themselves when buying wood: the species, the moisture content, the amount of wood, the degree of preparation, and finally, the price.

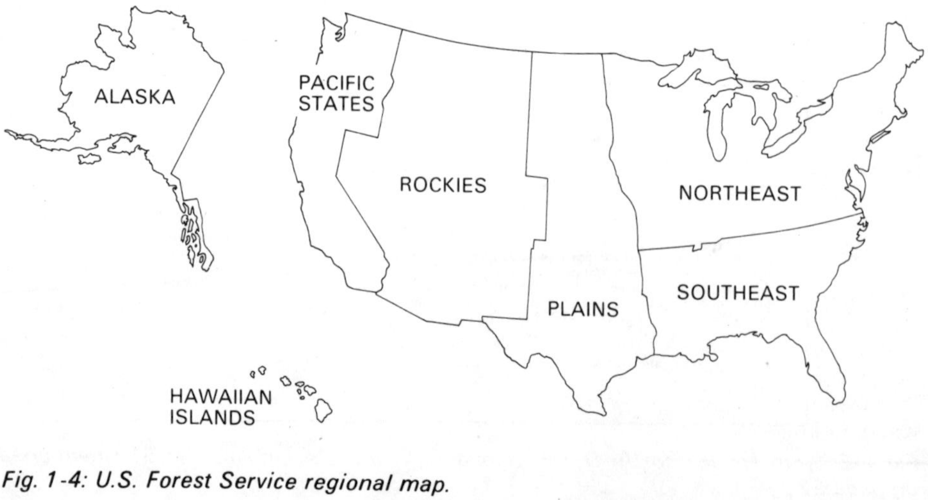

Fig. 1-4: U.S. Forest Service regional map.

Table 1-3: DOMINANT FUELWOOD TREES IN VARIOUS REGIONS OF THE UNITED STATES

Northeast (22 states)

Hardwoods	Softwoods
ashes	cedars
aspens	hemlock
beeches	pines
birches	spruces
cottonwood	
elms	
fruitwoods	
hawthorns	
hickories	
locusts	
maples	
poplar	

Southeast (9 states)

Hardwoods	Softwoods
ashes	southern pines
birches	
elms	
fruitwoods	
hawthorns	
hickories	
hornbeams	
locusts	
maples	
oaks	
poplar	
sweetgum	
sycamore	

Plains (6 states)

Hardwoods	Softwoods
ashes	cedars
aspens	pines
basswood	spruces
fruitwoods	
elms	
locusts	
oaks	
hackberry	
sweetgum	
sycamore	
maples	
hornbeam	
osage orange	

Rockies (8 states)

Hardwoods	Softwoods
hawthorns	pines
aspens	junipers
cottonwood	Douglas fir
alders	spruce
willows	firs
birches	larches
elms	hemlocks
oaks	

Pacific Northwest

Hardwoods	Softwoods
oaks	pines
alder	Douglas fir
maples	spruces
eucalyptus	larches
	firs
	juniper
	incense cedar

Alaska

Hardwoods	Softwoods
birches	western hemlock
aspen	Sitka spruce
balsam poplar	canoe cedar
red alder	yellow (Alaska) cedar
cottonwood	tamarack (larch)
willow	white spruce

Hawaii

Native	Exotic*
lehua	kukui (candlenut)
hala	paper mulberry
koa	pines
wiliwili	
naio	
kiawe	

*Refers to trees which have been introduced into the islands from abroad.

The first two considerations—the species and moisture content—have already been discussed. However, most wood dealers do not take the time to segregate the wood grades. It is usually cut and piled as it falls. As mentioned earlier, it is to your benefit to recognize firewood by the bark so you can determine the value of the wood. A good mix of wood grades is not a bad buy; it is handy to have some of the lower quality wood on hand for use as kindling and for heating on warmer days.

Wood is usually advertised as green or seasoned. Seasoned or dry wood will give you more heat and start burning more easily, so it is preferred. But, "seasoned" is a very vague term, so be sure and pin down how long it has been since the wood has been cut. It should have been seasoned at least six months. Many people buy their wood in the spring or early summer to be sure it will get the proper seasoning before winter, and because they are buying in the off season they usually get a lower price to boot.

Amount. It is important to know how much wood you are buying. As a general rule, the most common measurement of firewood volume is a cord. A *standard cord* measures 4 feet by 4 feet by 8 feet. But, all firewood measurement units (Fig. 1-5) include the air space between the sticks. Thus, the amount of solid wood depends upon whether the sticks are straight or crooked, round or split, and large or small in diameter. The variation is considerable as a standard cord may contain from 60 to 110 cubic feet of solid wood. A commonly used conversion from gross volume to solid wood content of hardwood sticks 3 to 8 inches in diameter is 80 cubic feet per standard cord. Larger diameters of round wood or split wood, neatly stacked, usually yield more solid wood per cord. A standard cord cut into shorter lengths will always stack in less space than originally because many of the crooks are eliminated and some wood is lost as sawdust. A cord of green wood will shrink at least 8 percent in volume during seasoning.

A *face cord* (also known as a *short cord* or *run cord*) is a pile of wood 8 feet long, 4 feet high, and as wide as the lengths of the wood cut, usually 12, 16, or 24

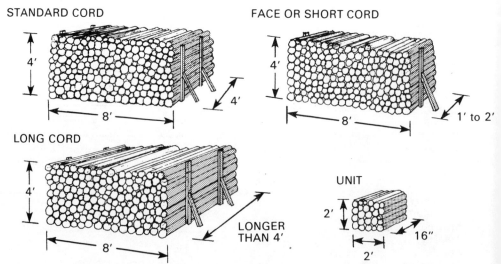

Fig. 1-5: Various firewood units of measurement. The law in many states requires fuelwood to be sold by the standard cord or a fraction of this unit only.

inches. A face cord is always less than a full cord and should be priced less. A face cord of 24-inch logs is half the volume of a full cord, a 16-inch face cord is one-third the volume, and a 12-inch face cord is only one-fourth the volume. A *long cord*, which contains logs more than 4 feet in length, is seldom used today.

Wood is sometimes sold as a "unit," which is usually 2 by 2 feet by 16 inches and fits into a car trunk or station wagon. This is one-twenty-fourth of a standard cord. Apple and other fruitwoods are sometimes sold this way.

In some areas, wood may be sold by weight, but this is generally not a common practice. If buying by weight, the purchaser should be aware that he may be paying for a considerable amount of water when the wood is not thoroughly air dried, which reduces the wood's effective heat. One can get some idea of whether wood is air dried by looking for large radiating cracks on the ends of the fuelwood pieces and by observing the degree of dark color on the ends. Generally, as mentioned earlier, the darker the color, the longer the wood has dried.

A practice that is becoming popular is for several neighborhood fuelwood users to purchase a load of pulpwood (trees of poor quality or small diameter). These loads range between three and eight cords. Pulpwood is generally cut green, so it may require eight to 12 months of air drying before being burned.

Firewood is sometimes sold by the truckload. Of course, the amount of wood in a "truckload" varies greatly depending upon the type of vehicle. A pickup truck with a bed 4 feet wide, 19 inches deep, and 8 feet long will hold one 16-inch face cord. A dump truck may hold up to four standard cords. Large pulpwood trucks with a wood rack will hold from six to nine standard cords. One certainly has to be wary of the fuelwood dealer who tries to tell the wood purchaser that a pickup load is a full standard cord. A cord of undried red oak will weigh approximately 2 tons, so neither the pickup truck nor your car truck has the capacity to haul a full cord.

Degree of Preparation

Preparing wood is labor intensive, and the price is largely dependent on how much work you are willing to do yourself. If you want to have your firewood split into fine pieces and stacked near your doorstep, expect to pay a premium price. If you want to go the cheapest route, as previously mentioned, cut the wood yourself from standing trees. Wood can be bought in almost any stage between these extremes (Table 1-4).

Table 1-4:
COMMON METHODS OF BUYING FIREWOOD

Method of Sale	User Labor Involved
Truckload of green logs	8 hrs./cord
Truckload of 4-foot rough split wood	6-8 hrs./cord
Cut and split, you pick up	1-2 hrs./cord
Cut, split and delivered	0
Stumpage	8-12 hrs./cord
Slabs, ends, etc.	Varies
Pallets & scrap wood	Varies

It is fairly common to buy wood in 4- or 16-foot lengths and do some of the cutting and splitting yourself. Wood delivered in this form usually costs about half that of wood cut and split, but involves about eight hours of hard work. There are no real "deals" in buying wood, just exchange of your sweat for the dealer's sweat. But, if you enjoy cutting and splitting wood as many people do, you can be paid handsomely for your recreation.

Price

Prices for the same types of wood vary from place to place according to the labor costs, transportation costs, the supply, and the demand. Shop around, try to buy when the demand is low and be sure you understand exactly what you are buying. Here are some pointers which will help you to be a wiser wood buyer.

1. Buy from a dealer who has established a reputation. Ask a wood burning friend or neighbor where he/she buys wood. You could check to see how long the dealer has been doing business.

2. Be wary of newspaper ads, particularly "special deals." Look for catch-phrases such as "truckload" or "fireplace cord," etc. The chances are they are delivering no more than one-third to one-half of a cord. That is alright if you know what you are going to get. Pay for no more than what is delivered.

3. If you are purchasing a large quantity of long logs, avoid buying those that have been skidded through mud and dirt; what you save in buying bulk can be lost in ruined saw chains. Also, the dirt, sand, and gravel cannot burn and will quickly clog your fireplace or stove.

4. Never buy unstacked wood; even an experienced eye can be fooled by a loose pile of wood. If the dealer refuses to stack the wood, go elsewhere.

5. Apply your knowledge for judging seasoned wood. Do not be hasty when buying; carefully appraise the wood as to seasoning, quantity, and quality.

6. Buy during periods of low demand. During the summer, most people are not thinking about cold snaps and firewood. Wood sales are slow, so the dealer is anxious to make sales. Prices go up in the fall. If you want to season your own wood, buy in winter or early spring. If you buy unseasoned wood in mid-summer you can forget about burning it until late winter unless it is split and placed in a good place to dry.

7. Do not be unreasonable. You have the right to expect seasoned wood in a standard quantity, but you cannot expect a dealer to have a full cord of perfectly matched shagbark hickory logs for you. Most dealers advertise and sell "mixed hardwoods," which contain a variety of the better known hardwoods. If these are well-seasoned, you are getting a good buy. If you do demand a cord of pure oak or hickory, be prepared to pay a nice premium for same.

Firewood Cutting Tools 2

It was Henry David Thoreau, America's first great nature writer, who said, "Wood warms you twice; once when you cut it, and again when you burn it." The statement is as true today as it was when Thoreau said it over 130 years ago. What is also true is that gathering and preparing wood for your wood burner, even taking into consideration modern work-saving equipment, is no picnic. Under the best of conditions, it involves plenty of work, and not a little sweat. Still, if you plan your work, acquire the right tools, and gain the knowledge of how to use them, you can reduce both the amount of work and the sweat. After all, it was the same Thoreau who also said, "A man needn't earn his living by the sweat of his brow, unless he sweats a lot easier than I do."

CHAIN SAWS

The chain saw is the Number 1 firewood cutting tool. It has replaced the hand crosscut saw, buck saw, and bowsaw, and it is steadily taking over many of the functions of the humble axe. In view of this, it is important that the "wood burning lumberjack," who is considering purchasing a chain saw or who already owns one, knows how to maintain and safely use it. To help you do this best, we recommend that you obtain a copy of the McCulloch publication—YOUR CHAIN SAW, The Complete Guide on How to Use Your Chain Saw—from your local chain saw dealer; or write to McCulloch Corporation, P.O. Box 276, New Ringgold, Pa. 17960 for details.

Competition between reputable chain saw manufacturers has become so keen that any chain saw which obtains an appreciable hold on the market today must be up to high performance standards. They are all designed and constructed

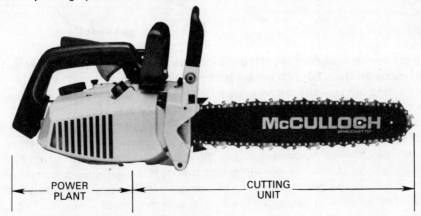

Fig. 2-1: The two working units of a chain saw.

to perform under various operating conditions. It would be a fairly safe bet that if there is a failure, it is in all probability due to neglect of one of the fundamental principles of reasonable care and attention.

The chain saw is basically a mechanical device comprising a power plant and a cutting unit (Fig. 2-1). These saws are produced in many types, shapes, and sizes. Manufacturers are constantly improving the already powerful engines by producing saws that are lighter in weight, develop more horsepower, and provide for faster chain speeds. The chain saw has gradually evolved from the heavy, expensive tool of the professional logger to a lightweight, low-priced tool available in a variety of stores. Their growth into world-wide use has been phenomenal, and chain saws are now virtually a household word, especially in wood burning homes.

Gasoline or Electric Chain Saws

Although a few chain saw models are electrically driven (Fig. 2-2), by far the most popular and useful for most firewood lumberjacks are those powered by an air-cooled, two-cycle gasoline engine. They eliminate the need for dragging around a lengthy extension cord, and they enable you to work anywhere that you can reach. Electric saws are less expensive, but your reach is limited by the length and capacity of the extension cord. Electric chain saws do not require fuel storage and are easier to start. Also, they do not produce exhaust fumes and, therefore, can be used for woodcutting and construction projects in the woodshed or basement. Electric chain saws employ the same chains and bars as smaller gasoline saws. They do not generally have the log- or tree-cutting capacity (diameter of tree) of gasoline powered saws.

Fig. 2-2: An electric chain saw is ideal for cutting firewood near a convenient power source.

Electric saws are generally rated in amperes of electric current. A typical small saw having an 8- or 10-inch bar usually has a motor that uses 8 to 10 amperes, while a medium size chain saw with a 12- to 14-inch bar will be equipped with a motor that requires about 10 to 12 amperes. They are designed for 110/120 volts AC. Be sure to check your owner's manual or refer to the chart on page 42 for the proper size extension cord to use with your electric chain saw.

Gasoline chain saw power is rather difficult to determine since there are, in most cases, no published horsepower ratings. However, cubic inch displacement of the engine cylinder (area of the bore times the length of the stroke) can be used as a rough guide, since the displacement sizes of engines are directly comparable and can be easily checked. The larger the displacement, the greater the power.

Firewood Cutting Tools 29

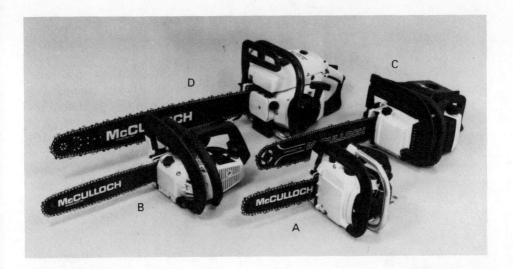

Fig. 2-3: A basic group of chain saws: (A) Mini-saw; (B) lightweight; (C) medium duty; and (D) professional model.

Many manufacturers classify their gasoline saws in groups or series (Fig. 2-3). Although there are no "official" standards, the basic groups are mini-saws, lightweight chain saws, medium-duty chain saws, and professional chain saws.

Mini-saws generally have an engine displacement of about 2.0 cubic inches and are equipped with 10-, 12-, or 14-inch guide bars and chains. Their weight (including the cutting attachment) is from about 8 to 12 pounds. Mini-saws are light, easy to handle, and inexpensive.

The lightweight chain saws usually have a displacement of 2.0 to 3.0 cubic inches and weigh between 10 to 14 pounds. While they are available in guide bar lengths from 12 to 20 inches, those in 14- to 16-inch lengths are the most popular.

Medium-duty chain saws generally have a displacement of 3.0 to 4.9 cubic inches and weigh between 12 and 20 pounds. Bar lengths range from 14 to 26 inches, with the 16- and 20-inch ones being the most popular. The medium-duty saw is best for the serious amateur woodcutter.

Professional models are used primarily by the logging industry and have engines with up to 9 cubic inches of displacement. They weigh from 15 to 40 pounds and have guide bars as long as 38 inches or more.

While the manner in which a gasoline two-cycle engine operates may be interesting, it is more important to understand the functions of several components of an engine (Fig. 2-4) which must work in unison if the chain saw is to run properly. These include the following:

Fuel Tank. The fuel tank is normally located within the power unit of the saw with the filler cap positioned on its top. Its function is to store the fuel mixture.

Fuel Filter. Located in the fuel tank, the fuel filter is connected to the line that runs from the tank to the carburetor. It filters particles from the fuel so that

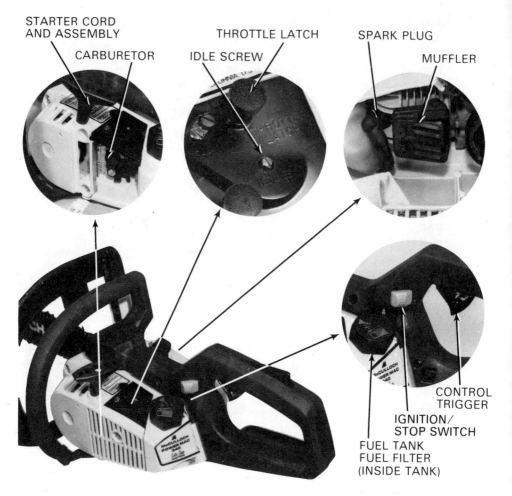

Fig. 2-4: The major components of the engine's fuel system.

they will not obstruct the fuel flow to any part of the fuel system. This filter is usually a replaceable unit.

Carburetor. The function of the carburetor is to mix fuel and air at the correct ratio. The carburetor is connected to the engine.

Idle Screw. This screw is an adjustable mechanical stop for the throttle, and it regulates the speed of the engine at idle.

Throttle Control System. The complete system includes the throttle control trigger; throttle control linkage; and, in some cases, a throttle control latch or throttle control lockout, or both. The control trigger is located in the handle so that the saw can be gripped with both hands and the control trigger can still be actuated or released. The throttle linkage itself transmits the operator's movements of the control trigger to the carburetor. The throttle latch locks the throttle partially open for easy starting.

Ignition/Stop Switch. The ignition/stop switch is designed as a safety device to turn the chain saw OFF. That is, the switch is so located that it can be moved to the OFF position by the operator while maintaining a grip on the handle, or handles, with both hands. It must be in the ON position to start the saw.

Starter Assembly. The starter cord and return spring are contained within a housing that is usually located on the left-hand side of the chain saw. The function of the starter assembly is to rotate the engine by cranking.

Spark Plug. The spark plug is located in the cylinder head and is used to ignite the fuel air mixture.

Muffler. The muffler on a chain saw reduces the engine noise levels. When the muffler is equipped with a spark arrester, it also greatly decreases the possibilities of fire. A spark arrester screen is required when operating a chain saw in some states and all federal forest lands.

The transmission of power from the chain saw engine to the cutting chain is accomplished through one of two types of drive—direct or gear. Gear drives are used mostly by professionals who use larger cutting chains and need to apply heavy cutting pressure. Direct-drive saws, on the other hand, are lighter, have fewer moving parts, are easier to use, and have a lower initial cost. Chain saws using this type of drive are popular because less operator fatigue is experienced during use due to their light weight and ease of portability. Little operator effort is required, as the sharp, fast moving chain tends to pull the saw down through the log.

The Cutting Unit

The cutting attachment is made up of three parts: a saw chain, a guide bar, and a sprocket (Fig. 2-5). On direct-drive saws, the saw chain is pulled around the guide bar by a sprocket attached to a clutch drum, which is driven by the chain saw engine.

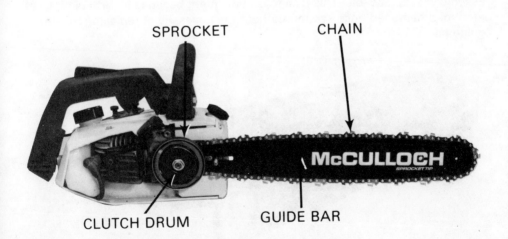

Fig. 2-5: The major components of the cutting unit.

Most chain saws use a centrifugal-type clutch which comprises small shoe elements attached to the crankshaft that rotate inside a small clutch drum as the crankshaft turns. When the engine is idling, the spring-tensioned shoe elements are held away from the drum face. An increase in engine speed increases the crankshaft rotation so that the tension on the shoe spring is overcome and the shoes engage the clutch drum, causing the drum to rotate and drive the chain sprocket. The clutch of a direct-drive chain saw will slip if the chain speed is reduced by applying too much heavy pressure while cutting or by binding of the chain thus preventing damage to the engine.

As for the chain drive sprocket, there are three basic types: floating rim, spur, and fixed rim. The floating rim sprocket mounts on a splined hub which is attached to the clutch drum and is free to move from side to side. It can align itself with the guide bar rails. The fixed rim sprocket is the same as the floating rim type except that it is permanently attached to the clutch drum and must be manually adjusted for proper alignment. Spur sprockets, similar in shape to a gear having wide teeth, support the saw chain either directly on the teeth tips or by nesting the chain drive links between the teeth. The floating rim sprocket is easy to replace, it has the floating alignment feature, and its reduced contact area minimizes chain wear.

The guide bar is the track that supports and guides the saw chain. Bars are manufactured in three basic types:

1. Hard tipped bars (Fig. 2-6A) have hard surface material welded on the nose to minimize wear and increase service life.

2. The standard sprocket tip (Fig. 2-6B) or roller nose bar has the same general configuration as hard tipped bars except that in the nose it has a toothed roller inserted which supports the chain, reduces the heat buildup caused by chain friction on the nose rail, and permits longer chain life.

3. The sprocket nose bar with replaceable nose (Fig. 2-6C) is similar to the standard sprocket tip bar. An added feature of this bar is that the complete nose assembly is replaceable. This design is convenient because it permits replacement of a damaged nose section without the necessity of replacing the entire guide bar.

Fig. 2-6: Three major types of noses for the guide bar: (A) hard-facing; (B) sprocket tip bar; and (C) replaceable sprocket nose bar.

Firewood Cutting Tools 33

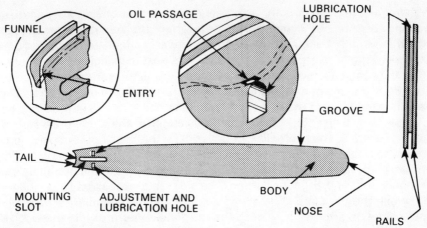

Fig. 2-7: Parts of a guide bar.

Guide bars are available with three groove widths: .050 inch, .058 inch, and .063 inch. These grooves are compatible with the three most common saw chain gauges or drive link thicknesses. The relationship of the outside diameter of the sprocket and width of the bar entry is extremely important because of the large number of saw chain drive links that must enter the guide bar when the saw is being used (Fig. 2-7). As the saw chain leaves the sprocket, it wants to continue in a circular path, and the stresses necessary to retain the chain on the guide bar rail are severe. Of course, the shape of the saw chain drive link as it enters the bar groove, the contour of the guide bar body, and the radius of the guide bar nose are also important features. Thus, it is important to follow manufacturer recommendations when replacing sprockets, guide bars, and saw chains to eliminate any reduction in cutting efficiency.

Guide bars are available in various lengths. It is best for the average firewood cutter to get a bar that will make it possible to do the majority of the cutting in one cut (Fig. 2-8). But, there are instances where you can use a two-cut method. A 12-inch guide bar, for example, can halve a 24-inch log in two passes.

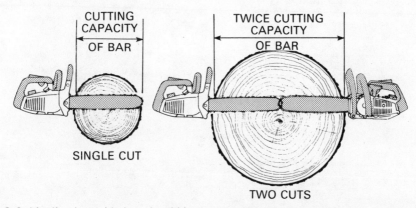

Fig. 2-8: Ideally, the guide bar should have a cutting capacity slightly larger than the diameter of the largest wood to be cut (left). Actually, a tree can be cut which has a diameter larger than the cutting capacity of the bar.

34 Firewood and Your Chain Saw

When selecting bar length, also keep in mind that chain saw engines can frequently be fitted with guide bars of several lengths. As a rule, the range of bar options is usually based on the engine's power, as well as balance. To avoid underpowering or overpowering the saw, it is best to select a guide bar length within the range as recommended by the manufacturer.

The most important part of a chain saw is the cutting chain. It is like a bicycle chain. It is made up of metal parts riveted together to form a jointed loop which is flexible lengthwise. The parts which make up the chain and their names are shown in Fig. 2-9. The center link, or drive link, of the chain hooks into the sprocket so that the chain can be driven around the bar. In addition, the saw chain has both left-hand and right-hand cutters. These cutters have a top plate that is sharpened to a fine edge for cutting. The side plate cutter cuts the wood on the side, while the top plate chips wood out of the middle similar to a chisel. The depth gauge determines how deep the top plate penetrates into the wood each time it cuts a chip.

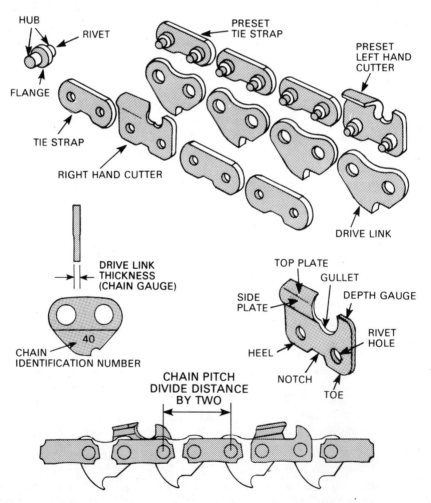

Fig. 2-9: The parts of a saw chain.

Like chain saws, there are numerous types and manufacturers of chain. To a large extent, cutting performance of the chain saw depends on the type of chain being used as well as the manner in which it is maintained. For instance, a good general purpose saw chain and the one most frequently furnished as standard chain with new lightweight chain saws is chipper chain or round chains (Fig. 2-10A). This is the chain that has made the chain saw popular today and has been proven during some 25 years of service. It is the least expensive of several chain types and is relatively easy to file and maintain. The chipper chain is not a high production chain but is available in most pitches and all standard gauges.

Another popular chain for lightweight saws is the semi-chisel or semi-square chisel (Fig. 2-10B), which has been in use for approximately 20 years. Its main feature is a small working corner which permits rapid cutting. It can easily be sharpened with a round file and requires no special filing technique. Semi-chisel chains are available in 3/8-inch pitch and all standard gauges.

One of the fastest cutting saw chains on the market is the chisel chain (Fig. 2-10C). The main feature of this chain is its square-shaped cutters, which sever the wood fibers with a single cutting pass. The unique square cornered cutting tooth can be sharpened with a conventional round file and is available in 3/8-inch pitch and all standard gauges.

Other styles of saw chains are available which are designed to be sharpened with an automatic sharpener. The cutter itself has a distinctive claw-like shape that sticks up to allow the grinding stone in the automatic sharpener to reach the cutting edges. In addition, the saw chain with a special guard or an antikickback design are also available.

Safety Options

As chain saws have become more popular, they have also become more varied. Not too many years ago, if you went to purchase a chain saw, you had a few basic brands to choose between. Having chosen the brand and the size, you had no more decisions to make. Today, however, it is a different story—there are a great many options from which you may select. The safety options should be given utmost consideration.

Kickback Protection Devices. Sometimes the chain, at the nose of the guide bar, digs into the wood in such a way that the saw violently kicks up and back. At other times, twigs may get caught in the chain or the chain may strike a foreign object during a cut. On other occasions, the top portion of the chain may

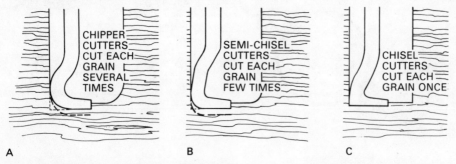

Fig. 2-10: Three types of cutter chains.

become pinched. All these actions cause a momentary stoppage of the chain and result in kickback (Fig. 2-11). Kickback can be very dangerous and is one of the most common causes of serious chain saw accidents.

Manufacturers have also helped to reduce the chances of kickback and kickback related accidents. By far, the best of the kickback protection devices is the chain brake. While it does not really prevent kickback, the chain brake is designed to stop the chain before it can do any real damage. If a kickback occurs, the operator's hand on the front handle of the saw (Fig. 2-12A) tends to hit the brake lever, triggering the device (Fig. 2-12B) and stopping the moving chain in milliseconds. Some manufacturers offer the chain brake as an option, while others feature it with all their saws.

Chain Catcher. Chain breakage is not a common occurrence, especially if you keep the chain well-sharpened. However, when a chain breaks, it could cause injury to the operator's unprotected hands or body.

Many saws incorporate a catcher pin near the base of the bar that is designed to help protect the rear hand by limiting the travel of a broken chain. As additional protection to the right hand, a few chain saws offer a right- or rear-hand chain guard.

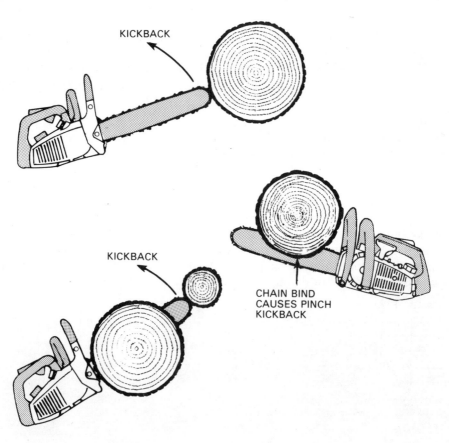

Fig. 2-11: Avoid kickback because it can be dangerous.

Firewood Cutting Tools 37

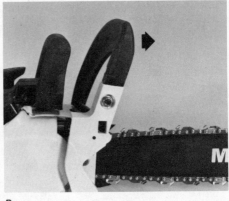

A   B

Fig. 2-12: (A) Position of the chain brake lever when the saw is operating normally. (B) Position of the chain brake lever when it is engaged.

Trigger Interlock. Another important safety device is the throttle interlock, located on the rear-hand grip of the saw, usually above and behind the throttle trigger. The interlock makes it very difficult for the throttle trigger to be opened accidentally or when the hand is not in the proper position.

Anti-Vibration System. If you have ever operated an old chain saw for any length of time, chances are that you felt your hands tingle after the engine was shut off. This was caused by the vibration of the engine and produced considerable fatigue and annoyance.

To overcome this difficulty, many saws are now available with some type of anti-vibration system. Most of these systems incorporate rubber cushions between the handles and the vibrating parts.

Muffler Shields. With some models of saws, the engine muffler is exposed, and since it can heat up to 800 degrees F, anyone touching the muffler accidentally could receive a severe burn. A muffler shield will, of course, help to reduce the chance of such burns. The one shown in Fig. 2-13 is an integral part of the clutch cover, which, incidentally, houses all chain brake components and diminishes any possibility of accidental contact with the muffler. It also helps to reduce the possibility of igniting any dry material, such as pine needles, while the saw is operated with the clutch side next to such combustible material.

Other Features

Although the safety features just described should be high on your list of priorities, other optional devices may also be important to you. Here are some of the more popular (and useful) features available.

Oilers. As the chain moves around the guide bar in a special groove, friction builds up heat. A lubricating oil must be pumped into the groove during cutting to combat this friction. The chain lubrication is accomplished by two methods, both supplied by the same oil reservoir located under the engine.

1. **Manual chain oiler.** The manual oiler provides lubrication to the chain when a button, usually located near the throttle, is pushed.

2. **Automatic chain oiler.** The automatic oiler provides a continuous flow

Fig. 2-13: Safety options to look for.

of lubricating oil to the bar and chain during operation. The amount of flow is adjustable on most saws.

Some manufacturers offer both automatic and manual oiling. This is an advantage if you will be cutting in very cold or hot weather or if you will be doing a great deal of heavy cutting. Under such conditions, any additional oil can be provided by the manual oiler. The manual oiler allows you to keep cutting when the automatic oiler malfunctions.

Automatic Sharpeners. Automatic chain sharpeners are handy (Fig. 2-14). When the chain requires sharpening, simply depress the sharpener button for a few seconds with the engine running at about half speed. Depressing the button brings a grinding stone into contact with the cutting edge of the specially designed chain, restoring the edge to near peak cutting efficiency. However, even with the regular use of the automatic sharpener, periodic touch-up filing or sharpening will be required.

Electronic Ignition. Because cutting can require frequent stopping of your chain saw (while moving from one spot to another, refueling, checking chain tension, etc.), easy starting is an important consideration. The standard system used for many years with most chain saws was high tension magneto ignition. While this system was fine when it was working properly, it did require servicing and constant adjustment, as well as frequent replacement of the points and condenser. This system is still in use in some saws.

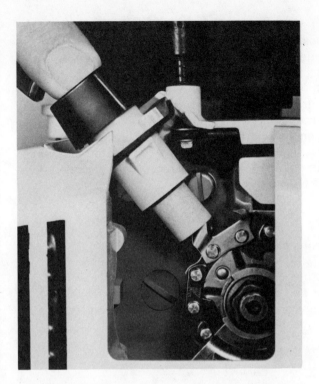

Fig. 2-14: The operation of the automatic sharpener.

Recently introduced to the chain saw field, the electronic ignition is a sealed system that does away with the condenser and the points. The ignition timing of the engine remains set for the life of the saw. Electronic ignition is ideal for the occasional user who would rather not do any tuneup work and replace parts.

It is important not to confuse electronic ignition with electric start. While push-button electric starting worked well for garden tractors and lawn mowers, it never proved successful with chain saws. As a result, it is not offered as an option by any of the major manufacturers.

Handles. In cold weather, metal handles draw heat from the operator's hands, resulting in quicker fatigue and discomfort. For this reason, most manufacturers are using either plastic or rubber-insulated handles, since they have low thermoconductivity when compared to painted steel or aluminum. In addition, plastic and rubber handles add a damping factor which reduces vibration. It is also wise to wear gloves when operating a chain saw.

Electric Saws. Electric saws furnish many of the same features that the gasoline type do (Fig. 2-15). Most electric chain saws are double-insulated for added safety. This means the unit is constructed throughout with two separate "layers" of electrical insulation. Chain saws built with this insulation system are not intended to be grounded. As a result, the extension cord used with a double-insulated saw can be plugged into any conventional 110/120 volt AC electrical outlet without concern for maintaining a ground connection. **Note:** Double insulation does not take the place of normal safety precautions when operating an electric saw. The insulation system is for added protection against injury resulting from a possible electrical insulation failure within the saw.

Other features and conveniences are offered by the manufacturers of electric or gas models that contribute, in varying degrees, to cutting efficiency and comfort. A "shopping tour" will familiarize you with what is available. Consider, too, another important "feature"—the manufacturer's experience and reputation as well as that of your dealer. When testing a chain saw, do not forget to examine it for quiet running, low vibration, handling ease, and balance. The latter two points are judged on "feel" and convenience of controls (on/off switch, throttle, oiler, and choke buttons). The following is a recap of the considerations to keep in mind when purchasing a saw:
- Safety features.
- Lightweight.
- Clean, compact styling.
- Good balance and handling characteristics.
- Convenient controls.
- Operator comfort.
- Reputable manufacturer and dealer.

If you insist on these minimum considerations, you cannot go wrong when purchasing a chain saw—either gasoline or electric powered.

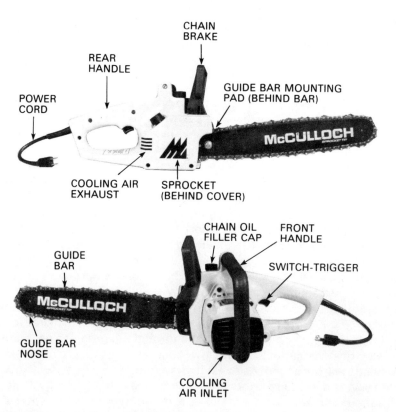

Fig. 2-15: Major parts of an electric chain saw.

Other Equipment

Other equipment can be divided into two groups: (1) for the operator and (2) for the chain saw. Let us first look at the items needed by the operator.

Personal Attire. The clothing worn by the chain saw operator should be selected to meet two requirements: safety (Fig. 2-16) and comfort. To achieve these two requirements, keep the following considerations in mind when selecting your clothing for chain saw work:

1. Wear trim-fitting garments that are neither too loose nor too tight. Loose clothing may be caught by the moving chain or may be drawn into the engine air intake. Also, such clothing may catch on branches or other projections and throw you off balance. Do not wear neckties, scarfs, or jewelry; wear cuffless pants and generally close-fitting clothes when running your chain saw. Clothing that is too tight may hamper your movement and agility.

2. Clothing should be suitable for the weather. Warm, but not bulky clothes are best for winter. Lightweight clothing is preferable in hot weather.

3. Always wear snug fitting, nonslip work gloves. They will improve your grip and protect your hands.

4. Protect your head with a hard hat when felling trees or working in the woods. These hats, similar to those worn by construction workers, protect against falling bark, dead branches, and other debris which may be dislodged from the tree overhead. But, hard hat protection is not needed for all chain saw work. If, for instance, you are bucking up firewood in the yard, you really do not need a hard hat.

5. Wear work shoes or boots, preferably with metal toe reinforcing. Low shoes or soft shoes should not be worn. Calked or hobnailed boots are excellent for working in the woods, on rough ground, or on top of logs. Nonskid soles should be worn when the footing is slippery.

6. Protective goggles or safety lens type glasses are a must while operating a chain saw.

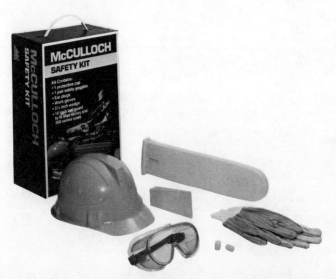

Fig. 2-16: Safety kits provide many of the items needed when using a chain saw.

7. Be sure to be fitted for and wear hearing protection devices (head set or ear plug types).

Chain Saw Accessory Equipment and Supplies. There are a few accessory items and supplies that you should have regardless of the chain saw operation you plan to undertake. For example, except in the work area, always keep a scabbard over the saw chain and guide bar, or keep the chain saw in a carrying case (Fig. 2-17). The latter encloses the entire saw.

To carry a supply of fuel mixture, use a red safety fuel can(s) with a filter nozzle. Also take along oil for the chain oiler and a funnel to fit the oiler reservoir. Of course, when using an electric chain saw, you will need a properly sized extension cord and a source of electrical power (Table 2-1). In the woods, a portable electric generator, capable of 1.2 to 2 kilowatts, is an ideal power source for most electric chain saws. The generator should be operated as directed in the manufacturer's instruction manual.

Table 2-1: EXTENSION CORDS

Ampere Ratings	Volts AC	Minimum Gauge Wire Length of Cord		
		25 feet	50 feet	100 feet
5-6	120	18 gauge	16 gauge	14 gauge
6-8	120	18 gauge	16 gauge	12 gauge
8-10	120	16 gauge	14 gauge	12 gauge
10-12	120	16 gauge	14 gauge	10 gauge

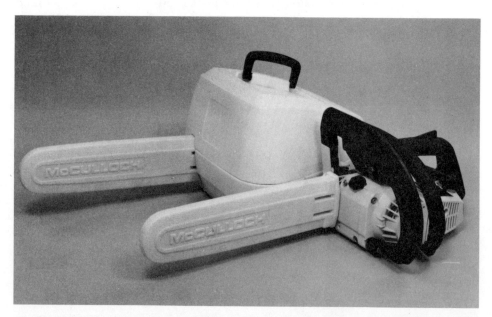

Fig. 2-17: When transporting the chain saw from place to place, be sure to carry it in a carrying case or at least cover the guide bar with a scabbard.

A few simple tools (screwdriver to fit the chain-tensioning screw, spark plug wrench, grease gun if required for roller-nose, chain sharpening kit, and wrench to loosen nuts on saw) may be needed for chain and saw maintenance. It is also good to have the following "spares" on hand: proper spark plug, air filters, and an extra loop of sharpened saw chain. When working in the woods under dry conditions, a fire extinguisher and shovel should be available in case of fire. A first aid kit should also be taken along. If cutting is to be done in an area where snakes are present, carry a snake bite kit. Snake boots and proper bite care are the best defenses against poisonous reptiles.

OTHER CUTTING TOOLS

Axes do have an important part to play in woodcutting. They are made in various patterns and head configurations (Fig. 2-18). Their heads are usually forged from carbon tool steel, and the blades or bits are heat-treated. Head weights vary from 1-1/4 to 5 pounds, with hickory handles ranging from 14 to 36 inches long. The double-bit axe is usually used to fell, trim, or prune trees and to split and cut wood. It is also used for notching and shaping logs and timbers. The single-bit axe may be used for the same purposes; in addition, the poll is used to drive wood stakes. Hatchets are used for cutting, splitting, trimming, and hewing; nails and stakes may be driven with the striking face.

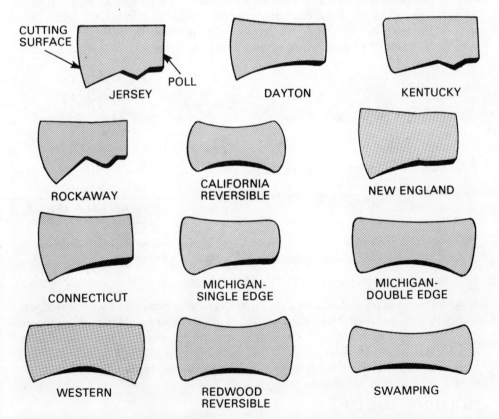

Fig. 2-18: The basic axe head shapes.

Axe Care. A good axe will last almost indefinitely if given proper care. Never leave your axe lying on the ground. Not only is this practice dangerous, but ground moisture will quickly rust the axe head and cause the handle to rot. When not in use, secure your axe by imbedding it into a block of wood or a stump, or hang it up. If you are using your axe in winter, warm up the head before starting to chop. Even finest steel can get brittle at extremely cold temperatures (below 20 degrees F). In any case, a warm axe will cut better. If you are not going to use your axe for some time, rub the head with a preservative oil, and hang it up in a dry place.

For both efficiency and safety, axe blades must be kept sharp. Even with moderate use, periodic resharpening is required. Place the head in a vise and run an angled file from the heel to the toe over the entire length of the blade. Turn the axe frequently and concentrate on forming an even edge over the entire blade. After the nicks have been removed, use a circular stone to hone with a circular motion at about a 25-degree angle. First use the medium side and then the fine side of the stone. An Arkansas or moon stone will perfect and polish the edge. Always use plenty of sharpening oil when using a stone, as it keeps the stone's surface from clogging with steel particles.

The Axe Handle. The axe handle takes a lot of punishment. The best ones are made of hickory, which is very hard wood and is highly resistant to splitting (Fig. 2-19). Most handles are treated with oils at the factory to retard rotting and to add to the handle's resiliency. Some people like to wrap friction tape around the handle where it is gripped to insure a firmer hold. Never abuse the handle by using it as a hammer or as a base for pounding.

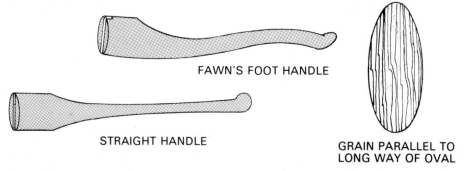

FAWN'S FOOT HANDLE

STRAIGHT HANDLE

GRAIN PARALLEL TO LONG WAY OF OVAL

Fig. 2-19: The axe handle, showing two popular shapes. Hickory is the favorite wood, and the grain should run in the direction of the long axis of the oval.

When purchasing an axe, make sure that the head is secured by visible wedges. Never buy an axe with the head molded to the handle by plastic, as replacing the handle will be difficult. When replacing the handle, buy the best quality hickory available. An inferior wood will not only split or crack sooner, but will vibrate in your hands, making chopping painful. When replacing the head, secure it with iron or hardwood wedges (Fig. 2-20).

HOLDING TOOLS

The sawbuck is the simplest way to hold logs when cutting them to firewood

Firewood Cutting Tools 45

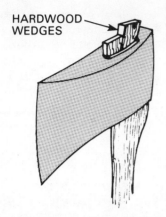
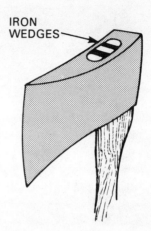

Fig. 2-20: Installing a new axe handle. Hardwood wedges are placed parallel to the long dimension of the head; iron wedges are placed perpendicular to the same dimension.

lengths. As shown in Fig. 2-21, sawbucks, frequently referred to as "horses," can be made in several different styles.

    Sawbucks are not difficult to make, and no special tools or materials are required. Hardwoods and dried logs make the longest lasting sawbucks, but there is no reason why you cannot make a fine sawbuck out of white pine or green logs. For that matter, scrap lumber will do fine if it is in reasonable condition. To make the all-log sawbuck, select a log about 4 inches in diameter. Cut two 2-foot and four 3-foot pieces. (The length of the longer pieces may vary to suit your best cutting height.) Assemble the longer pieces into a pair of X's, using spikes, nuts and bolts, or lag screws. Leave about 1 foot above the intersection. Then, adjust the two X's to exactly 90 degrees, and fasten them together using the short pieces. The other two sawbucks are made in a similar fashion, except

Fig. 2-21: The sawbuck or horse, shown in different styles, using different materials. Sturdiness is more important than style or material.

46 Firewood and Your Chain Saw

Fig. 2-22: The two-legged sawbuck.

that they are longer (about 4 feet). Remember to adjust the X's to exactly 90 degrees. Spikes are best for all-lumber sawbucks, while lag screws or spikes are good for the log/lumber sawbuck. It should take no longer than one-half hour to build a good sawbuck.

For field cutting operations, use a two-legged sawbuck such as shown in Fig. 2-22. It is held together with a bolt and a wingnut or with rawhide lashes, and it can be carried into the woods conveniently. Another good field log holder is a log lifter, such as the one shown in Fig. 2-23. This device is also handy for moving logs about without using too much muscle.

Fig. 2-23: A log lifter in use.

SPLITTING TOOLS

Wood splitting is a necessary step in the preparation of firewood for consumption by either a stove or fireplace. It can be done by hand or by using a power splitter.

Hand Splitting Tools

The cutting axe described earlier in the chapter can be used to split logs. The

Firewood Cutting Tools 47

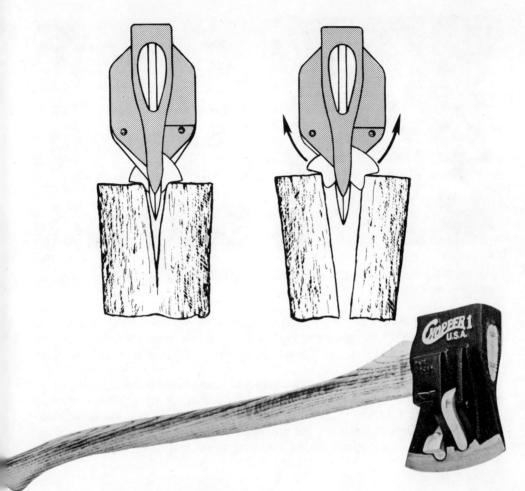

Fig. 2-24: Typical special log-splitting axe.

job, however, is a lot easier if you employ either a woodchopper's maul or a splitting axe. The latter is especially designed just for splitting logs. Figure 2-24 shows how a typical splitting axe works. The downward force of the special axe causes the splitting levers to contact and rotate. This rotation transfers the downward force to a direct outward force, which splits the log wide open. The splitting levers also prevent the blade from sticking in the log. If the first stroke fails to split the log, just raise it and split it with the second stroke.

The woodchopper's maul is designed especially for splitting wood (Fig. 2-25A). It is also used in conjunction with wood-splitting wedges (Fig. 2-25B), first for making a notch with the splitting edge, and then to drive the wedge with the striking face opposite the splitting edge. Mauls are forged from high-carbon steel, are heat-treated, and usually are made in 6- and 10-pound head weights with approximately 32-inch hardwood handles.

The correct maul strike is an accurate one, not a forceful one. Mauls can power their way through even the most stubborn logs without much help from you.

48   Firewood and Your Chain Saw

Fig. 2-25: (A) Using a woodchopper's maul for splitting a log; (B) using a maul with a wedge to split a log.

Misplaced maul swings are often very unforgiving on the tool's handle. The great head weight and swing momentum make split or shattered maul handles more of a possibility than with axes. So be extra cautious and precise when swinging a maul.

The wood-splitting wedge is usually forged from a solid piece of high-carbon steel and may be heat-treated. Wedges are made in various patterns, such as the ones illustrated in Fig. 2-26. In addition to the three popular types—the square-head, Oregon splitting, and the stave wedges—there is the new conical-shaped wedge. The conical design allows it to seek lines of least resistance

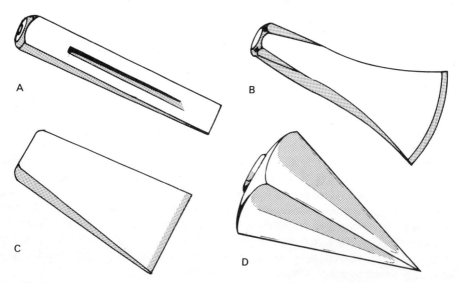

Fig. 2-26: Four major types of wedges: (A) square head; (B) Oregon; (C) stave; and (D) conical shaped.

Firewood Cutting Tools 49

Fig. 2-27: Typical four-way splitting wedge.

when driven into a log. Regardless of type, most wedge weights range from 2 to 8 pounds. Always use a woodchopper's maul or an axe to make a starting notch. A wedge should be struck with a sledge or woodchopper's maul having a striking face that is larger than the head of the wedge. Do not use plastic or aluminum wedges described in Chapter 3 when chain sawing, because they will not survive the rigors of wood splitting. There are special splitting wedges, such as the so-called four-way types (Fig. 2-27), that work well on larger logs.

Power Splitters

Power splitters do take a great deal of the work out of splitting. They do the work much faster, too. Therefore, if you plan to do a large amount of cutting—and splitting—wood, it may be wise for you to consider the purchase of a power splitter. With a splitter, the work involved in splitting a log includes the lifting and setting of the log on the splitter, moving away, and starting the splitting process. During the operation of the splitter, keep hands and arms away from the machine. There are two types of power splitters: the hydraulic-wedge type and the screw or cone type.

Hydraulic Splitter. With this type of splitter, the log to be split is rested on a metal beam or pan, and a hydraulic cylinder then pushes the log into a steel wedge to split the wood. Designed to handle 24-inch or longer logs, they can be coupled to the hydraulic system of a farm tractor or can be driven by their own power source—gasoline or electric (Fig. 2-28). Actually, most log splitting done by the do-it-yourself woodcutter is also done on home grounds. This is the reason for the great popularity of hydraulic electric log splitters (Fig. 2-29). Most, if not all, seasoning and storage of firewood is done at home.

Fig. 2-28: Hydraulic log splitters can be driven: (A) by a tractor; (B) by their own gasoline engines; or (C) by electricity.

Some of the large hydraulic models generate a 10-ton or better splitting force. For those who wish to provide the power behind a hydraulic log splitter, there are hand-driven models available (Fig. 2-30).

Rotary Screw or Cone Splitters. With this type of splitter, the log is pushed up to the cone, which screws its way through the log until the log splits. These units (Fig. 2-31) are available with their own power source, can be mounted on

Firewood Cutting Tools 51

Fig. 2-29: While electric log splitters are most popular for backyard use, they can be used in conjunction with a generator for splitting in the woods.

Fig. 2-30: Hand-driven hydraulic log splitter.

the rear wheel of an automobile or garden tractor, or driven by a farm tractor power take-off (PTO). On wood with interlocking fibers, the screw or cone splitters frequently have difficulty in splitting the wood cleanly. And, in some cases, an axe or maul may have to be used to complete the job.

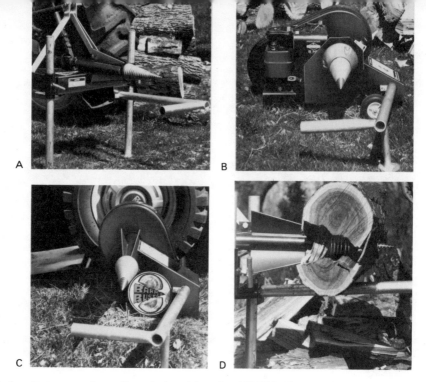

Fig. 2-31: A rotary screw log splitter being driven by: (A) PTO; (B) gasoline engine; and (C) auto drive. (D) The rotary screw log splitter in action.

Woodchopper's Block

While not often thought of as a tool, the woodchopper's block is essential to anyone attempting to split wood with hand tools. Blocks make a steady base for the log to be split and prevent wedges, axes, and mauls from going into the ground where they may be damaged by dirt and stones. Almost any flat stump will do as a splitting block. A good block should be at least double the diameter of the majority of the logs you plan to split (Fig. 2-32). Logs have a tendency to jump about upon impact, and a broader block will provide the room it needs to move without falling off. Block height will vary, but the best height is one that will allow you to make initial impact at about mid-thigh level. This causes least strain on the back and makes for more efficient splitting. The best blocks come from hardwoods that are hard to split and that are highly seasoned. If you have such a block, it is a good idea to protect it just as much as your good tools.

Fig. 2-32: The woodchopper's block holds the log steady and prevents tools from being pounded into the dirt.

Wood for Your Woodpile   3

As stated in Chapter 1, the least expensive way to obtain a supply of wood is to cut it yourself. There are various places to gather wood. If there is not any wood on your property or if you do not have friends with trees to spare, it is possible to purchase "stumpage" rights from farmers and rural property owners. Permits can frequently be obtained to cut wood on lands owned by municipal, state, or federal government agencies. State and federal foresters periodically fell diseased trees and offer the wood to those willing to haul it away. Local and state highway departments also offer wood if you are willing to relieve them of it (especially after a damaging storm). Power and telephone companies often cut swaths through wooden areas and have no use for the wood they cut. And, if you are fortunate enough to live in logging country, sawmills and lumberyards usually have slabwood and waste woods (Fig. 3-1) that they give away or sell for a minimal fee.

Fig. 3-1: "Leftovers," such as these cull logs, are often made available by timber companies to firewood buying prospects.

City dwellers, too, can have access to free wood. For instance, as much as 30 percent of the contents of dumps and landfills is usable firewood, free for the asking. Drive around your area right after the next wind storm. You are bound to see lots of blown over trees whose owners would probably be glad for you to cart them away. Every year in parks or along city streets, trees die, and all you need to remove them is permission from your local park or street department. Also, be on the lookout for old orchards and other more or less urban stands of trees about to be displaced by new construction.

Let us now take a closer look at all these sources of wood and see how you can add to your woodpile.

Woodlots

If you or a friend owns a woodlot, it is possible to expect a harvest of anywhere from one-half to one cord of wood per acre each year, depending upon the quantity and quality of the trees. Cutting trees from a woodlot for firewood can result in double benefits. It can result in both an economical fuel for home heating and in an improved woodlot where growing potential for higher valued trees has been increased. It is often the common practice to cut the straight, well-formed trees for firewood because they split easier than their crooked, limby neighbors. This type of cutting practice rapidly reduces woodlots to stands of inferior trees having no market value for years to come. With quality timber values what they are today, it is unwise to follow such a practice.

Cutting firewood, if done properly, can improve a timber stand by removing those trees that will never make valuable crop trees or that are interfering with the growth of the better trees in the stand. With careful selection of trees harvested for fuelwood, the remaining trees will grow more vigorously because they will have more soil moisture, nutrients, and sunlight. Usually, the trees that should be cut are the poorly formed and the lowest value trees in the woodlot. A firewood harvest is much like a culling or thinning operation (Fig. 3-2).

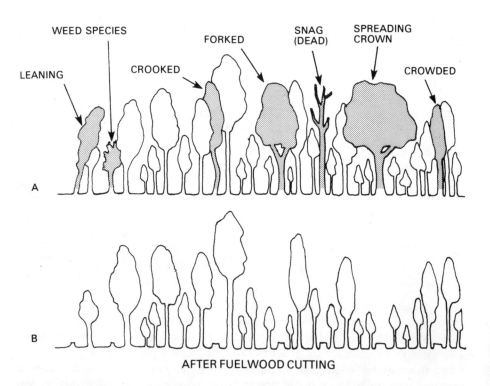

Fig. 3-2: Before and after thinning a woodlot. "A" shows a profile of a stand of trees that has not been thinned. Shaded trees are those to be harvested. "B" shows the same stand after shaded trees have been cut. Remaining trees will grow at a faster rate, insuring future fuelwood.

Woodlot improvement may necessitate the removal of good quality trees that are crowding equally good trees or better ones, all competing for limited sunlight and soil nutrients. This is a higher form of timber stand improvement and requires good judgment. Professional state service foresters, consulting foresters, and extension foresters are available to assist forest landowners in making these decisions. These foresters are prepared to advise landowners on forest management, timber harvesting, and marketing procedures. The names and locations of these individuals are available through county Cooperative Extension Service offices.

There is one important consideration often forgotten when cutting firewood: Be sure to respect the wildlife residents in the woodlot area. Before cutting any tree just for its fuel value, try to determine whether or not that tree has cavities that serve as active dens for squirrels, songbirds, and other wildlife. Forest nut and fruit trees—oaks, hickories, hop hornbeam, flowering dogwood, and so on—are important to squirrels, deer, turkeys, bears and other animals. By allowing some of these den and food trees to remain in your woodlot stand, you will be amply rewarded by having continued wildlife activity.

Also, before thinning a woodlot, you should know the value of the trees with which you are working. Oaks and hickories make the best firewood, but they are slow-growing trees. Sugar maple, ash, white (paper) and yellow birch are more valuable than red maple, beech, elm, or aspen (poplar). If you are uncertain as to species, your local forester or county agent, as previously mentioned, is in a position to assist you and offer advice as to woodlot thinning. The best time to begin the thinning process is when the trees average between 4 to 10 inches in diameter at breast height—about 4-1/2 feet above ground. Trees are growing rapidly at this stage, and prudent thinning will add more heating value to the trees left standing, while the culls will be of use in your wood burner.

Rural Sources

You would be surprised to know the amount of wood lying about that could be yours for the asking. Many farmers have woodlots that they either care little about or have little time to care for. Be sure to approach the farmer (or owner, if he/she is an absentee) to obtain permission first. Both of you should be certain as to which trees are to be harvested and how much is to be taken. It is best to get it in writing, but this is not absolutely necessary. If the farmer uses a wood heating system, you might volunteer to leave the farmer a portion of your wood for his own use. Some owners might charge a small stumpage fee, but this is usually minimal. **Caution:** Avoid damaging property (fences, roadways, buildings, etc.) in the process of gathering your wood, and always clean away debris. Leave the woodlot looking better for your having been there. Once you have established good relations with the farmer, you probably have secured a source of firewood for years.

Commercial Logging. Not to be overlooked as a source for fuelwood are wastes and leftovers from commercial logging. Slash materials—tops, limbs, and other unusable parts of the tree—are frequently left lying in place. Removing this slash actually serves to protect the forest, as dried waste materials have been the cause of many forest fires. Also, many trees are knocked over in the process of felling and removing trees. Many logging operators permit private individuals to remove this waste for fuel. Since this material is of small size, it is ideal for a stove or fireplace. The very small stock makes excellent kindling, and it is all easily handled by an individual weekend logger.

Sawmill Operations. Once logs reach the sawmill, more wastes, often in huge quantities, are created. Slabs, trim, edging, and leftover sections are useless for lumber, but they make excellent fuelwood. Many sawmill operators or lumber companies offer these waste materials at very reasonable prices simply to reduce their own hauling and disposal expenses. Also, much of this wood has been down for some time, so it is already fairly well aged. In addition, it is already cut up, so your splitting problems are eliminated.

Diseased Trees. Disease, unfortunately, kills many trees. Disease has wiped out the chestnut tree, has decimated the American elm, and threatens the white pine. The chestnut and the white pine may be harvested for firewood, but elms which have died as a result of the Dutch Elm disease present a problem. The trees must be removed because they are an eyesore, they are a danger to people because of falling limbs, and they continue to spread the disease. Some suggest that these trees be cut and burned. Elm is just an average wood for heating purposes, and it is difficult to split, but it does form good coals. Other people are opposed to burning elms which have died as a result of the Dutch Elm disease because they believe that the disease is spread by burning the wood. Experts differ as to the propriety of burning this wood. Some communities have passed ordinances against burning such trees. Be sure to clear everything with local foresters or officials before burning diseased elm trees. Actually, such trees might as well be burned because there is no acceptable alternative; burying or dumping spreads the disease, too.

Public Lands

One of the richest sources of free firewood is the national forests and areas administered by the federal Bureau of Land Management (BLM). Due to the energy crisis, many of these public forests have recently been opened to wood cleaning. The type of wood offered includes cull logs and slash resulting from commercial logging operations, windfall limbs and trees, deadwood, diseased trees, smaller trees cut during thinning operations designed to improve the growth of timber left standing, and trees removed during the construction of roads and firebreaks. Regulations generally allow each family to cut and gather a specific amount of firewood (normally 10 cords) for home use at no charge or a minimal fee for a given period of time (usually two weeks to several months). Wood cutting permits are required, however, and before one will be issued, the applicant must comply with prescribed regulations. For instance, public forest rules usually require all chain saws to have a spark arrester screen on the muffler. Additionally, you must carry a shovel and fire extinguisher.

The permits are issued only at the forest ranger district or BLM area where wood cutting is being allowed. Before applying for a permit, be sure to write or call the specific national forest headquarters or BLM district office for the latest cutting information and regulations.

Many state parks and forestlands have also opened their doors to selective thinning by weekend wood cutters. Downed timber or deadwood can be gathered easily, and in most cases, forest managers mark the standing trees they wish to have removed. State foresters, county extension agents, as well as county and city foresters can also provide you with information on local sources of firewood.

Other Sources

Much land is cleared every year for housing, shopping malls, highways, etc. All too often any trees knocked down in the process are simply removed to the local dump to be buried or burned. By inquiring of the proper authorities, you may be able to remove the trees at your own expense. Keep your eyes open for areas being cleared. Also, read the local newspapers; they often run accounts of land to be cleared long before the actual clearing starts. The name of the builder or contractor might be mentioned. Do not overlook any opportunities along this line.

Utilities. Public utilities, in the process of installing telephone lines, high tension wires, underground cables, etc., frequently must remove trees prior to installation processes. By contacting the local office, you may be able to obtain firewood. The telephone company is a very good source of firewood, since they must replace telephone poles at regular intervals due to deterioration. Poles are excellent fuelwood as they are well-aged. If the part of the pole that was in the ground has rotted away or has been treated with creosote, as is usually the case, do not attempt to burn that portion. You will get little heat from the rotted wood, and it may produce an objectionable odor.

Natural Disasters. Trees can be severely damaged or killed by storms, floods, lightning, and landslides, or they may simply topple over. By contacting the proper sources, you may be able to obtain a good supply of fuel for your wood burner—and it is free. Other trees simply die on the stump and turn into ugly snags. In most cases, the owner, whether a public or private party, will be more than happy to let you remove the tree(s). Snags make excellent fuel since the wood is already aged (dried out) when you cut down the tree.

Driftwood. If you live near any body of water—ocean, lake, pond, river, or stream—do not overlook the chance to pick up driftwood washed ashore. Because driftwood usually has been dead for some time, it is usually dry, even if it has just been taken from the water. Just let the pieces dry in the sun for a few days, and it is ready to burn. Avoid waterlogged pieces as they will never burn. It is best to burn driftwood only in a fireplace; the odd shapes do not readily accommodate themselves to the confines of a stove. However, do not expect a great amount of heat to be produced from driftwood, and do not start a fire with it.

Dumps and Landfills. Most downed trees sooner or later wind up in the town dump, since many local ordinances forbid open burning. Handling and disposing of this bulky material runs up municipal costs, as it takes up space sorely needed for solid wastes. You will need official permission to remove burnable wood, since many dumps sell salvage rights. The quality of these woods, however, may vary.

Many less obvious sources of firewood are overlooked. At construction sites, there is usually a lot of scrap material that is frequently thrown into the trash cans to be carted away. There is no reason why this wood cannot be used in your wood burner. Supermarkets get supplies in wood crates which are then discarded. Warehouses and other commercial operations use hundreds of pallets which are simply thrown away after use. It is possible to obtain a cord or more just from these pallets. More scrap lumber is simply thrown into alleys. Junk furniture may be useless for its original purpose, but excellent for the fireplace.

Also, look for demolition projects, frequently making available tons of excellent fuelwood. Some apartment owners simply dump scrap lumber and discarded wood items right on the street. The sanitation department is glad to have you remove them. The same department may have scrap wood it is willing to give away or sell at extremely low rates. Sometimes newspapers carry ads for free or cheap fuelwood, usually in the form of scrap. Trees along streets die or have to be removed for street widening; the highway department may be looking for someone to remove them and save itself hauling and disposal charges. The parks division may have dead or diseased trees that they need to get rid of; more fuelwood for you. Try placing a want-ad in the local newspaper, you might be surprised at the results. Landscapers, nurserymen, lawn and tree service businesses all remove wood from private property and public lands, which they are more than happy to sell at very reasonable prices to get something back for their scrap.

CUTTING FUELWOOD

Before building up your wood supply, plan how to do it safely and efficiently. What trees are to be cut? What handling will be necessary to transport, buck, split, season, and store them after they are felled? The answers to these questions depend upon your available trees, size of piece which should be burned, time available for seasoning, and equipment available for handling. In most cases, it is best to buck the trees in the woods to their final stove or fireplace length and then transport them home. Remember, the key to efficient fuelwood production is minimal handling.

Cutting firewood is hard work, but the chain saw, if used properly, can make it a great deal easier. To accomplish this, you must approach the operation of your chain saw with attentiveness and respect. According to a recent Consumer Products Safety Commission survey, 35 percent of all accidents to casual operators and helpers are **caused by inadvertent contact with the moving chain,** while 23 percent are **caused by kickback.** A careless move, such as reaching across or holding the work near the moving chain, or loss of footing and subsequent loss of saw control, account for many accidents.

In a book of this size and content, it is impossible to give full details on the operation of the chain saw, its maintenance, and its use to fell, buck, and limb trees. We suggest that you carefully study your chain saw owner's manual and read a book such as the previously mentioned McCulloch publication—YOUR CHAIN SAW. However, in this chapter, we will cover some of the important points to remember when cutting fuelwood. For example, keep these basic safety rules in mind:

1. Know your chain saw. Read the owner's manual very carefully. Learn the saw's applications and limitations, as well as the specific potential hazards. Do not attempt operations beyond your capacity or experience.

2. Always wear buttoned shirt cuffs and generally close-fitting clothes when you run your chain saw, to avoid the possibility of clothing getting caught in the saw. In other words, wear the proper apparel: safety footwear, snug-fitting clothes, hard hat (in wooded areas), safety goggles or lenses, hearing protection device, and gloves.

3. Do not use any other fuel than that recommended in the owner's manual. Refuel in a safe place. Open fuel cap slowly to release any pressure which may have formed in the fuel tank. Do not start a saw where you fuel it; move at least 10 feet from the fueling area before starting. Do not overfill or spill fuel. If fuel has been spilled on the unit, be certain the saw has dried before starting it. Do not refuel a hot saw—allow it to cool off. Also, never smoke while fueling or operating the saw. Keep a fire extinguisher or shovel handy.

4. Before starting an electric chain saw, be sure that the extension cord and connections are in good condition. Never operate an electric saw where wet ground or damp foliage can cause electric shock. When cutting, keep the cord clear of the chain.

5. Be sure that any helpers or spectators are at a safe distance from you and the saw, and that they are not standing where they might be struck by falling branches, etc. Keep bystanders from the work area. Remember that the operation of a saw should be restricted to mature, properly instructed individuals.

6. Start your saw without help. Do not hold a saw on your leg or knee. Keep all parts of your body and clothing away from the saw chain when starting or running the engine. Before you start the engine, make sure the saw chain is not contacting anything. Never operate a chain saw when you are fatigued.

7. Hold the saw firmly with both hands when engine is running; use a firm grip with thumbs and fingers encircling the chain saw handles and watch carefully what you cut. Be sure not to let the end of the chain at the nose of the guide bar hit branches, stubs, stumps, or any object other than the one you are cutting. Inattention in holding or guiding the saw while cutting can cause kickback.

8. Only make cuts within the capacity of your chain saw. Stand with your weight evenly distributed on both feet for proper balance. Do not cut in awkward positions (off balance, outstretched arms, one-handed, etc.). It is recommended that you **do not** operate a saw while in a tree, on a ladder, or on any other unstable surface. If you elect to do so, be advised that these positions are extremely dangerous.

9. Clear away brush, rocks, or anything else in the working area which might hinder your movements. Use extreme caution when cutting small brush and saplings because slender material may catch the saw chain and be whipped toward you or it may pull you off balance. When cutting a limb that is under tension, be alert for springback so that you will not be struck when the tension is released.

10. Make sure that the saw chain is moving at full speed just before it enters the wood. Early revving causes the engine to race too much; late revving will cause binding and clutch slippage. Keep the cutting speed under careful control. Modern chain saws cut rapidly. It is very easy to cut too deeply or at a wrong angle. For best control, hold the bumper spikes in contact with the wood as the chain begins cutting.

11. Do not stand in line with the bar and chain. Learn to stand to one side of the cut while sawing. Never touch or try to stop a moving chain with your hand.

12. Before cutting any standing tree, prepare an escape route from it at an angle of 45 degrees in the opposite direction from the intended line of fall of the tree. Always make an undercut or notch and never cut a standing tree completely through. The "hinge" is necessary to control the direction of the fall of the tree. Use plastic or aluminum wedges to control the fall of a tree or prevent

binding during bucking. Do not, however, use hard metal wedges or an axe to hold cuts open. Do not fell a tree during high or changing winds.

13. Study the effects of bucking and limbing cuts on logs which may roll. Always cut and work on the uphill side of a log. Also, be sure to limb with your feet and legs in the clear and try to keep the trunk of a felled tree between you and the limb being cut (Fig. 3-3). Do not limb with the nose of the guide bar; kickback can result which can prove very dangerous.

Fig. 3-3: When limbing, it is best to keep the trunk between you and the guide bar (left). In cases where this is not possible (right), exercise care to make sure that you do not allow the saw to swing into your leg.

14. Never touch or let your hand come in contact with a hot muffler, spark arrester, or a spark plug wire. Never run the saw with a fuel cap loose or without a muffler, exhaust stack, or a spark arrester. Keep the screens and baffles clean.

15. Avoid prolonged operation of your chain saw and rest periodically, especially if your hands or arms start to have a loss of feeling, swell, or become difficult to move. These conditions can reduce your ability to control a saw. Also, never operate your chain saw in confined or poorly vented areas.

16. Turn off your saw when moving between cuts and before setting it down. It is best to carry or transport a shut-off saw with its bar scabbard or guard on. When this is not practical, you should carry the saw with the guide bar and saw chain to the rear and the muffler away from your body.

17. Observe all local fire prevention regulations. Do not operate a chain saw when the weather is extremely dry and there is a fire hazard. Normally, authorities close forests to logging operations when such conditions exist. It is a good idea to stay in the cutting area for at least 15 minutes after stopping work to be sure there are no smoldering embers in the area. Put out any fires and report them, listing causes, if known, to the proper authorities.

18. Do not allow dirt, fuel, or sawdust to build up on the engine or outside of the saw. Keep all screws and fasteners tight. Never operate a chain saw that is damaged, improperly adjusted, or is not completely and securely assembled. Be sure that the saw chain stops moving when the throttle control trigger is released and the saw returns to idle. Keep the handles dry, clean, and free of oil or fuel mixture.

19. Never operate a saw with a loose chain. Check the chain tension frequently. Always stop the engine when adjusting the chain tension. Also, make sure that the chain and bar are receiving enough oil.

20. Keep the chain sharp. Touch up the chain teeth every couple of hours or whenever the sawing starts producing sawdust rather than chips of uniform size. It is important to keep in mind that in addition to tiring you, a dull chain can lead to excessive pressure on the saw bar and accidents. But, when you are sharpening the chain or adjusting the guide bar, wear gloves or take extra precautions not to draw your finger across a saw tooth. Both the sharpness of the teeth and their shape make them inflict cuts easily.

Operation of a Chain Saw

To become familiar with a new chain saw, it is a good idea to practice holding it and simulating a work stance without the machine running. Once you feel comfortable handling the saw, start the engine and cut some wood.

Holding the Saw. Wear nonslip gloves for maximum grip and protection. Grasp the front handle bar firmly with the left hand so that your fingers wrap around it, keeping the handle bar diameter in the webbing between your index finger and thumb (Fig. 3-4). Grasping the handle bar in this way gives you maximum control of the saw and reduces the chances of your hand slipping into the moving chain. Your right hand should wrap around the throttle control handle in such a manner as to provide good saw balance. Always keep both hands on the saw. Never shift hand positions or cross arms for easing strain or for better reach. If your arms tire, stop the saw and rest for a while.

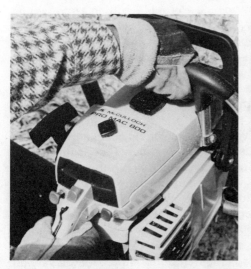

Fig. 3-4: Proper way to hold a chain saw and position oneself when cutting.

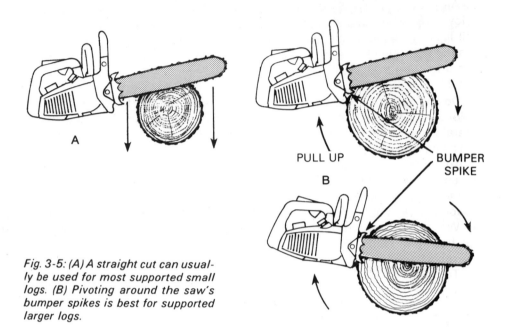

Fig. 3-5: (A) A straight cut can usually be used for most supported small logs. (B) Pivoting around the saw's bumper spikes is best for supported larger logs.

The proper stance is important, too. Clear away brush, rocks, or anything else in the working area which might hinder your movements. Be sure to have safe, sure footing, and always keep your weight as equally balanced as possible on both feet. Hold the saw so that the chain is not in line with your body. Always cut with your left arm extended as straight as possible. Since you will be exerting downward pressure to cut, guard against the loss of balance by being ready to hold up on the saw as it cuts through the wood.

Cutting Wood. The most desirable way to hold a log while cutting it into lengths is to use a sawbuck (see Chapter 2). When this is not possible, the log should be raised and supported by the limbs or logs. But, be sure that it is well supported. Then, start the chain saw; and after it is warmed and idling smoothly, you are ready to cut, keeping the following points in mind:

1. Holding the saw firmly in both hands and with the engine idling, bring the cutting unit up above the log, with the nose of the guide bar pointing slightly upward. Slowly lower the power unit portion of the saw so that the bumper spikes grab into the wood and act as a pivot; the chain should not touch the wood. If the saw has a manual oiler, lubricate the chain and bar. Then, squeeze the throttle trigger so that the engine is going at top speed (full throttle) as the chain touches the wood. Never run the engine slowly at the start or during the cut. Cutting at partial speed will allow the clutch to slip and burn, wearing it out early or giving a quick glaze on the clutch friction surfaces, which leads to even more slipping and burning.

2. Guide the saw, without forcing it, through the cut. Use only enough pressure to keep the chain cutting full wood chips. In other words, let the saw do most of the work.

3. Always cut as close to the engine end of the guide bar as possible. Also, employ the saw bumper spike when practical to act as a pivot during a cut. While small logs can usually be cut straight through (Fig. 3-5), larger ones are best sawn using the bumper spikes as a pivot. While the saw is cutting, be sure the chain and bar are being lubricated. A manual oiler should be pumped every 15 to 20 seconds during the cut.

4. Do not twist the guide bar, and **make sure that the chain at the nose of the bar does not touch anything.** Never allow a running saw to contact the ground or metal; one such contact can dull the chain more than cutting dozens of trees. Also, avoid knots, if possible.

5. To maintain complete control of the saw when nearing the end of a cut, ease up on the cutting pressure without relaxing your grip on the handles. Be ready to release the throttle trigger the instant that the chain breaks through the wood. Do not permit the saw to run at full throttle without a cutting load.

6. When cutting a log on a slope, always stand on firm ground, uphill and away from the area where the log might roll. Make sure no one is below where you are working.

7. To prevent the two sections of a log from coming together and binding or pinching the chain or guide bar when the cut is nearly completed, the cut must be made to relieve any stress. As shown in Fig. 3-6A, make a cut one-third of the way through and underneath the log using the top portion of the guide bar. Then, finish the cut from the top by sawing downward into the first cut using the bottom portion of the guide bar. The log will fall away from the guide bar.

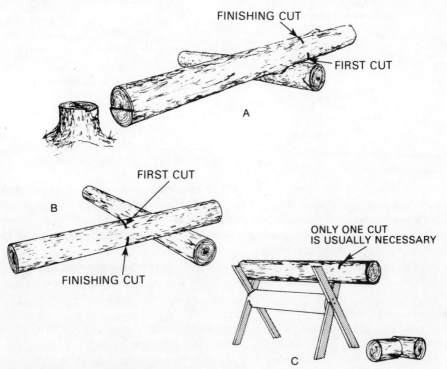

Fig. 3-6: Methods of cutting logs.

As shown in Fig. 3-6B, cut through one-third of the distance from the top using the bottom portion of the guide bar (pulling chain), and then finish from the bottom using the top portion of the guide bar. When small logs are supported on a sawbuck, as shown in Fig. 3-6C, cut all the way through from the top using the bottom portion of the guide bar.

8. Although the chain saw does most of the work, you must remain alert when using it. Cutting with a chain saw when you are overly tired is one of the worst things you can do. Take a break anytime you feel the slightest bit of fatigue.

Reducing Chain Saw Hazards. A significant number of chain saw accidents are caused by kickback. Kickback is the sudden backward or upward motion (or both) of the guide bar occurring when the saw chain near the nose or the top area of the guide bar contacts any object, such as a branch or log, or when the wood closes in and pinches the saw chain in the cut. For instance, when the chain on the upper 90-degree quadrant of the guide bar nose digs into the wood, the reaction forces the saw in a backward rotation towards the operator (Fig. 3-7). Pinch-kickback occurs when the wood closes in and pinches the chain on

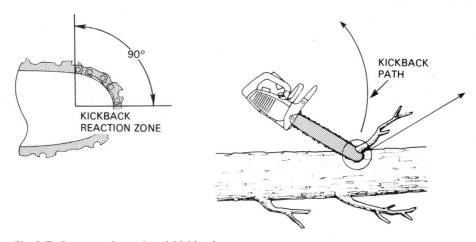

Fig. 3-7: Beware of rotational-kickback.

the top of the guide bar, causing a sudden backward thrust of the chain saw towards the operator (Fig. 3-8A). When making bucking cuts, with the chain on the bottom portion of the guide bar, the saw tends to pull away from the operator (Fig. 3-8B). The chain saw operator should avoid the following situations to reduce the probability of kickback.

- Abrupt change of wood characteristics (i.e., green to dry, knots, etc.).
- Running the saw too slowly, especially at the beginning of a cut or when boring.
- Buildup of damp sawdust.
- A twig caught in the chain and jamming against the work.
- Twisting of the saw so that the cutters dig into the wood.
- Hitting the chain at the nose of the guide bar against a solid object.
- Closing of the kerf (or cut) which pinches the chain.

Wood for Your Woodpile 65

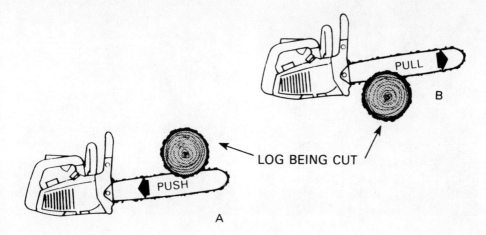

Fig. 3-8: The push (pinch-kickback) and pull reactions.

● A chain that is not tensioned properly, is dull, or has too low a depth gauge setting.

Maintaining a good balance is of the utmost importance when resisting any pulling, pushing, or rotational forces. Make sure that your stance is correct, with the body weight as evenly distributed on both feet as possible.

There are other chain saw hazards to guard against. For instance, a broken chain can be dangerous. However, if you position yourself so that no part of your body is in line with the chain, chances are very good that if the chain should break, it will run off the saw and onto the ground. To guard against the rare times when a broken chain will occur, most manufacturers have provided such safety devices as chain-catching pins and right-hand guards (see Chapter 2).

There is nothing complicated about chain saw safety. It is just the application of common sense to a handy, lightweight piece of machinery used to speed and ease the work of cutting wood.

Felling, Limbing, and Bucking

If you are a weekend lumberjack who uses the chain saw primarily for cutting firewood, you may not become involved in tree felling, since most likely you will be cutting dead and down timber. However, sometime you may attempt to fell a tree. Just remember that felling, limbing, and bucking take special skills, require utmost caution, and necessitate the use of safety precautions.

Small trees can usually be felled in any direction. With large trees, however, especially those that lean or have heavy branches on one side, you may have little choice in which direction they fall. That is, a tree may lean or be unbalanced due to uneven top growth or breakage even though the trunk does not lean. Large diameter branches are good indicators of imbalance. Prevailing wind direction affects tree growth and balance while present wind conditions affect fall direction. The inexperienced operator should attempt to fell trees only under conditions which indicate a high degree of certainty as to which way the tree will fall. Plan an exit route before starting.

66   *Firewood and Your Chain Saw*

Fig. 3-9: Felling a tree is accomplished with three cuts.

The actual process of felling a tree is accomplished with three cuts (Fig. 3-9). The face of the tree, or the direction in which the tree will fall, requires a horizontal cut and a sloping or angled cut of at least 45 degrees. The horizontal cut, which is made first, should have a depth of approximately one-third the diameter of the tree. These two cuts are called the undercut.

Make the felling or backcut about 2 inches higher than the horizontal cut. The felling cut should be kept parallel with the horizontal notching cut. Keep the guide bar in the middle of the cut so the cutters returning in the top groove do not recut the wood. Never twist the guide bar in the groove. Guide the saw into the tree—do not force it. The rate of feed will depend on the size and type of tree.

An inch or two of uncut wood should be left between the undercut and backcut. This acts as a hinge to hold the tree in line as it falls (Fig. 3-10). As the tree begins to fall, stop the saw engine and move back from the stump at a 45-degree angle, which will put you in a safe position should the tree roll or kick backward.

Wood for Your Woodpile 67

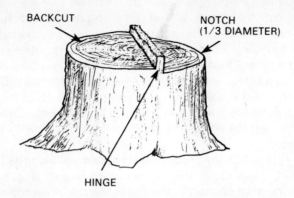

Fig. 3-10: The features of this stump indicate a properly felled tree.

A binding saw and closing kerf indicate an error in judgment. At the first such indication, remove the saw. If the saw cannot be removed, do not struggle with it. Shut off the engine, clear the area and plan a course of action using wedges to remove the saw (Fig. 3-11). Use only aluminum or plastic wedges—steel or iron wedges may damage the chain.

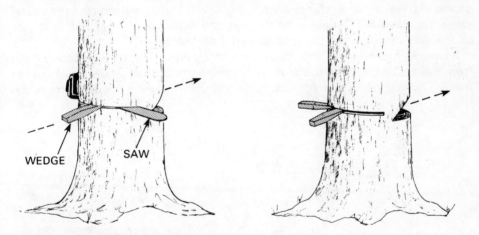

Fig. 3-11: Proper placement of wedges.

 Limbing a downed tree is not a difficult task, especially if it is done with a lightweight chain saw with a short bar. However, there are precautions that *must* be taken. For instance, before starting to limb up a felled or downed tree, check to be sure that some of the limbs are not holding the tree from rolling. Also, check overhead for broken limbs or chunks that may fall at any time.
 It is best to start at the butt of the tree and work towards the top so that the branches will be pointing away from you while you are working. Always stay on the uphill side if there is a chance the tree may roll or shift. Make sure you have good footing and a well-balanced stance. Clear limbs from your working area as you progress. For the most part, limbs are cut with the lower portion

of the bar, so beware of potential kickback. On limbs that are standing out free from the trunk, cut the bottom side of the limb one-third of the way through with a pushing chain, then cut down from the top to finish the cut with a pulling chain. When a branch is supported at both ends (by the tree at one end, and the ground at the other), topcut the branch first to prevent binding. Then, finish with an undercut. A wedge undercut is generally not necessary. On larger limbs, leave some connecting wood to guide or control the limb as it is falling away, and be careful it does not fall on you. To make bucking easier under some conditions, it may be best to leave some supporting limbs uncut. These may be cut off after bucking.

To buck a log that is on level ground and is supported along its full length with no apparent bind, make the following sequence of cuts (Fig. 3-12A). Set the bumper spikes and start the cut at the top of the log. Pivot the saw forward to make the backside cut. Stop the cut when the bar tip is still a few inches from the ground. Next, draw the saw back and drop down on the near face; then, dog in for the next cut. Always release the throttle trigger when the chain is running free and not cutting. Pivot the saw forward again as in the first cut and repeat this process until the log is nearly cut through. To finish the cut, throttle back to slow the chain and withdraw the bar until only the bottom portion of the tip remains in the cut. Move the tip back and forth across the uncut wood and continue to slow the chain speed as the cut nears completion. If there is thick bark on the log, watch for a color change in the sawdust that will indicate the sound wood has been cut. Release the throttle and withdraw the bar before it cuts all the way through into the ground and dulls the chain.

If a log is supported on both ends, as mentioned earlier, a single straight cut from the top would bind the saw, so examine the proper sequence of cuts to make. The first cut is an overbuck made through one-third the diameter of the

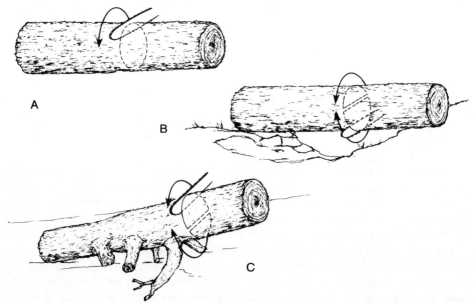

Fig. 3-12: How to buck a log under various stress situations.

log to avoid splintering. The second cut, an underbuck, meets the first cut and avoids pinching (Fig. 3-12B).

Until now, most of our cuts have been made with the bottom edge of the bar. This type of cut pulls the saw away from you and actually helps pull the saw into the log. When using the top portion of the bar, the chain saw will be pushed or thrusted towards you. Make sure that you have good footing and are able to stabilize yourself adequately. Be extremely alert to the possibility of a kickback. Occasionally, the cutters on the chain will bind when cutting like this, and the bar will kick up and back suddenly and severely. Be aware of this possibility, and at all times, maintain a secure hold on your saw.

When a log is supported on only one end (Fig. 3-12C), the first cut is made from the bottom up (underbuck). This cut should come up approximately one-third of the diameter of the log. As previously discussed, make sure of your footing and maintain a firm grip on the saw. Next, cut the back side of the log, and then make the final cut from the top down. By doing this, you can position yourself out of danger if the log should snap off unexpectedly. The end of the log should drop away and not pinch the bar. As a precaution against possible binding of the bar, again position your plastic or wood wedges in the cut as soon as possible.

When handling timber, it is never really safe until it is deposited as stove or fireplace ashes. You can never be sure, but you can exercise caution by knowing and respecting your chain saw, paying attention to safe work habits, and most important of all, by keeping your mind on your job.

GETTING YOUR FIREWOOD HOME

Once you cut the logs, you must get the firewood home. Traveling a hundred miles and back for a truckload of wood certainly will not save you any money. If you do not own a vehicle capable of hauling a good-sized load in one trip, team up with a friend or two and rent a truck or heavy-duty trailer. Lightly-sprung cars pulling regular duty trailers work fine for hauling small loads short distances on good roads, but even when dry, a cord of wood can weigh as much as two tons, and the rough roads of a forest can easily damage all but the toughest equipment.

Since you are going to drive on public highways, make sure your car, trailer, and hitch conform to state regulations. State police may issue citations for trailers and hitches that do not meet state regulations or for overloaded trailers. The car may require extra equipment, such as outside mirrors (right and left) before it is considered safe to pull a loaded trailer. The hitch should have safety chains to reduce sway and to back up the ball hitch. Drive well below the posted speed limits (they are for cars without trailers). Do not forget that the extra weight will add substantially to the distance required to bring the car and trailer to a full stop. **Caution:** If the roads are wet, icy, snow or slush covered, the stopping distance required may be doubled, tripled, or quadrupled. Stay to the right; do not obstruct faster moving traffic.

As mentioned earlier, rather than running the risk of ruining your car, when carrying a small load of logs, it is a much better idea to borrow or rent a pickup truck (at least one-half ton) to bring home the firewood. Pickups eliminate trailers and hitches and are usually built to handle heavier loads. They are usually built with standard equipment so that they meet traffic regulations. Make sure that the tires are in good condition—you are going to be going in places where just the finest tires (snow tires in winter) are going to be acceptable. Make sure that you have enough wood to fill up the truck. If you are going to carry more than the rated load capacity of the truck, the carrying capacity may sometimes be extended by installing slats on the inside of the box. **Caution:** The practice of overloading trucks is dangerous and is not recommended. Heavy-duty shocks and springs and oversized tires may extend the safety factors. Check with local police to determine whether a truck so extended may be in violation of state traffic regulations.

One of the major problems when bringing logs out of the woods is finding a convenient stacking or loading point. It is usually no problem to cut access trails suitable for a heavy-duty truck or jeep by felling small trees or pruning back overhanging limbs, but any number of obstructions may make it impossible to pull fallen logs directly to these access trails or a storage point.

While a four-wheel drive jeep or truck equipped with a front-mounted winch (Fig. 3-13) is ideal for solving this problem, the use of a snatch block or a block-and-tackle and several lengths of 1/2-inch or thicker rope can serve the same purpose. Simply attach the pulley to a convenient tree and pull the line in a direction which will move the log around the obstruction (Fig. 3-14). When there is open passage to the access trail and the pulling vehicle must move along at an angle to this route, a snatch block positioned at the side of the trail solves the problem.

Fig. 3-13: A front-mounted winch or a winch set in the back of a truck is handy for pulling out logs.

Wood for Your Woodpile

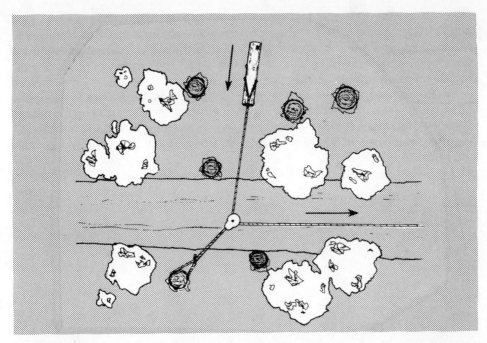

Fig. 3-14: Method of using a pulley or snatch block to skid a log in confined areas.

Once you have maneuvered the logs out onto the access trail, a logging sled or skid pan is the best way to transport the logs to the stacking or loading site. Construction of a simple skidding sled is no major task. Cut out the wooden runners on a band saw, and use sturdy pieces of 4 inch by 4 inch lumber for the cross beams. Raise the cross beams well off the runners so they will clear

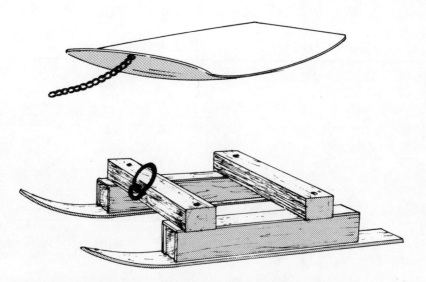

Fig. 3-15: Typical log skid pan and sled for small logs.

small obstructions. The pulling rope or chain can be attached to the heavy ring bolt fastened in the center of the front cross beam (Fig. 3-15). Use a peavey to roll the logs up a ramp of short poles and onto the sled. Fasten them securely with rope or chain, keeping as much of the log off the ground as possible. This reduces friction and dirt accumulation.

Skid pans, flat steel pans with rounded-up front ends, serve the same purpose as skidding sleds. The pans should be chained as close as possible to the pulling vehicle. This raises the pan's front end, again reducing friction and drag. Remember, you can easily strain all but the strongest vehicles if you try to move too much at one time.

Seasoning Firewood

It is one thing to acquire firewood, either by purchasing or cutting one's own supply, but quite another thing to get it ready to stoke into your stove or fireplace. As has already been mentioned in Chapter 1, burning green wood not only reduces its heating value by as much as 50 percent, but it also creates smoke which may be hazardous to your health and to your flue (because of the creosote formed) as well. Table 4-1 details the effects of seasoning of hardwoods on moisture and heat.

Table 4-1:
EFFECTS OF SEASONING OF HARDWOODS ON MOISTURE AND HEAT

	Moisture Content	Relative Heat Value
Green in fall, winter, or spring	80%	50 to 60%
Green in summer	65%	70 to 75%
Trees leaf-filled in summer after 2 weeks	45%	82%
Spring wood seasoned 3 months	35%	90%
Spring wood seasoned 6 months	30%	95%
Wood seasoned 12 months	20 to 25%	100%

There are a *few* occasions when it is desirable to burn green, unseasoned logs. For example, there are times when a bright, but not particularly hot, fire is called for; a green log(s) is good for such a situation. Also, an unseasoned log may be used as a moderator on a fire that has grown too energetic.

If you suddenly run out of seasoned wood during a rough winter, it is sometimes necessary to use green logs. In such emergencies, some woods are better to burn than others. The following woods, when green, will provide suitable heat: ash, beech, black locust, yellow and white birch, Douglas fir, larch (tamarack), osage orange, shagbark hickory, sugar maple, lodgepole pine, red and white spruce, Norway pine, and black cherry. But, whenever you burn green logs, be sure to frequently check the stovepipe and/or chimney flue for creosote (see Chapter 8).

There are three steps in seasoning or preparing fuelwood for the stove or fireplace: (1) cutting the logs to size; (2) splitting them to expose more wood surface to the air for faster drying; and (3) stacking them so they can air dry properly. Let us take a look at each of these steps in detail.

CUTTING THE LOGS TO SIZE

As a general rule, it is best to limb and buck a tree as soon as it is felled. The possible exception to this rule is when a tree has all of its leaves. Then, it is often best, if possible, to wait before cutting off the branches. Even though the tree is down, sap is still flowing and the leaves will remove much of the moisture through evaporation from their veins. When the leaves are thoroughly limp and beginning to turn brown, they have removed all the moisture they can. The tree can then be limbed and bucked into logs that are easily transported.

If you cut your own wood, the logs are usually bucked to stove or fireplace size in your backyard rather than out in the field or woods. If you purchase your firewood, it is generally cheaper to purchase 4- or 6-foot lengths from your wood dealer and cut them to size yourself. The cutting should be done on a sawbuck (Fig. 4-1) as described in Chapter 3, or in a cutting crib, such as the ones shown in Fig. 4-2. Cutting cribs make it possible to cut many logs at the same time. The logs can be piled together, and one pass of the chain saw will cut them all to precisely the same length. Normal stove lengths are 16 to 24 inches, while fireplace lengths generally range from 24 to 36 inches.

Fig. 4-1: Cut logs to fireplace or stove lengths.

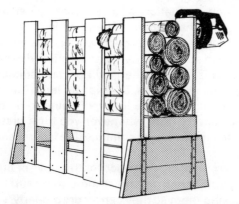 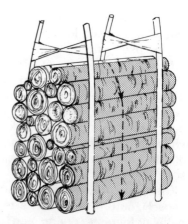

Fig. 4-2: Holding cribs such as these help to speed up the cutting of logs to size.

A question frequently asked is, "How long will it take to cut a cord of wood using a chain saw?" The answer is not a simple one because there are so many variables—the skill of the operator, the species of wood, the size and type of chain saw, the rate of work, etc. An independent testing company obtained the results shown in Table 4-2 under the following conditions: standard cord of wood (approximately 80 cubic foot volume), cord well stacked, logs limbed and cut to 48 inches, logs 6 to 10 inches in diameter, and chain saw with 14-inch guide bar.

Table 4-2
CORD CUTTING TIME

Test #1. If the logs are cut into 24-inch lengths:

Gasoline Saw
    Steady Rate: 2—3 hours
    With Breaks: 3—4 hours

Electric Saw
    Steady Rate: 3—4 hours
    With Breaks: 4.5—8 hours

Test #2. If the logs are cut into 16-inch lengths:

Gasoline Saw
    Steady Rate: 4—6 hours
    With Breaks: 6—8 hours

Electric Saw
    Steady Rate: 6—8 hours
    With Breaks: 9—16 hours

Note: The above times are based upon the safe use of the chain saw in accordance with manufacturer's instructions and recognized proper use of the chain saw. These times also assume that the chain saw is in good working order and allows for no time lost due to breakdowns. Using a cutting crib will reduce cutting time.

SPLITTING FIREWOOD

As already mentioned, the more surface of the cut wood exposed to the air, the quicker the firewood will dry and be ready for use. Therefore, it is important that the 4- to 8-foot transportable logs cut in the field or purchased from a dealer be cut to fireplace or stove lengths. Log lengths more than 6 inches in diameter should also be split so that drying is facilitated, more edges are exposed to speed up wood ignition and burning, and the pieces are easier to handle. Generally, if the logs are 6 inches or more in diameter, split them in half; if they are 12 inches or more in diameter, split them into quarters. In both cases, you will expose anywhere from 100 to 400 percent more wood to the open air (Fig. 4-3). An exception to the 6-inch splitting rule is the birch variety. Actually, any birch over 3 inches in diameter should be split. The birches have especially tight and impervious bark, preventing moisture from escaping. If birch logs are not split, they can decay (rot) in a year, even if placed in a shelter.

Wood splitting, as mentioned previously, can be done by hand or by using a power splitter.

Splitting Wood By Hand

When done the right way and with the proper tools, splitting wood by hand can be both fun and healthy exercise. Done improperly, it can be a backbreaking, dangerous task.

The hand splitting tools—axes, mauls, and wedges—and how to use them are fully described in Chapter 2. It takes practice to learn to split wood safely and

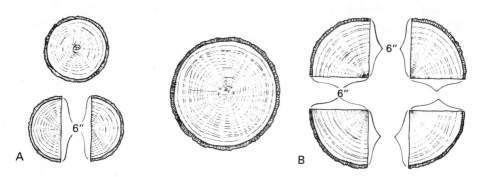

Fig. 4-3: The effects of splitting on exposing more wood surface to the air. (A) The bracketed portions of the log are now exposed to the air, more than doubling the amount of wood that can be dried. (B) All eight bracketed portions are exposed to the air, quadrupling the amount of wood that can dry at the same time.

efficiently. As detailed in Chapter 1 (see Table 1-2), different species of wood vary greatly as to splitting characteristics. Softwoods usually split easier than hardwoods and fruitwoods. Since straight-grained, knot-free wood is easier to split than crooked-grained, trees with many limbs have more knotty wood that is harder to split than trees with few limbs.

Some tips on splitting are:
1. Split parallel to spiral grain.
2. Short length logs split better than long ones.
3. Split parallel to knots.
4. Split in line with checks.
5. Green wood splits easier than dry, well-seasoned wood.

With either an axe or a splitting maul, a smooth, powerful splitting stroke is important. A good stroke has a telltale sound as it strikes the wood; it feels good and splits the wood cleanly. To accomplish a clean split, set the log (billet) on a solid base. (A chopping block such as detailed in Chapter 2 is ideal; *never* split wood directly on the ground as this is a good way to strike a rock or your foot.) Draw an imaginary line across the log's face in line with a natural crack, if any, or with its center. Drive the axe or maul in on that line as many times as it takes to form a split line across the face. Spread your legs slightly, extend arms straight from the body with the log squarely in front (Fig. 4-4A). When making the swing, flex your knees and follow through so that the angle between the log and your maul or axe is 90 degrees. Remember to keep your eye on the split line at all times. Sometimes on larger or tough-to-split logs, it is best to aim for the far side on the first swing (Fig. 4-4B). Then, make each successive swing closer to yourself. On easy-to-split logs, or logs partially split, aim to the side nearest you (Fig. 4-4C). This helps to avoid nicks and cuts on the maul handle.

If the log does not split, drive a wedge in at the center (Fig. 4-5A) or in a medullary ray or crack beyond the center of the log (Fig. 4-5B). If the first wedge does not split the log, or if it no longer drives easily, a second wedge should be started along a near-side medullary ray, so that the split will bisect the log (Fig. 4-5C). Some large logs of interlocking fiber woods may require three or more wedges to split.

Seasoning Firewood 77

Fig. 4-4: Splitting a log with a maul.

As mentioned earlier, frozen wood often splits quite easily, although it may be more difficult to pound wedges into it. This wedge problem can usually be solved by cutting a starting notch into the log face with your chain saw.

Logs less than a foot in diameter are usually halved as just described. Larger diameters are generally "quartered" (split in half, then split again to make four fireplace- or stove-size pieces). Larger logs can also be "slabbed."

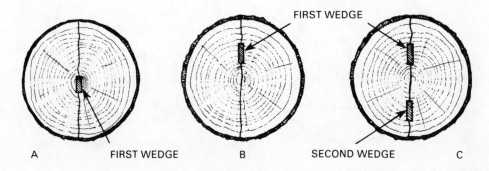

Fig. 4-5: Splitting a log using a wedge or wedges.

Slabbing is accomplished by splitting off the first piece of wood from the outer edge of a log section parallel to the bark (Fig. 4-6A), and then working a spiral pattern around the outside and in toward the center (Fig. 4-6B). Soon you will have a pile of slabs lying around the core, or heartwood, of the tree. You may find it easier to burn the heartwood whole rather than splitting it, since this wood, especially in old trees, often seems as hard as stone and only half as willing to split in two.

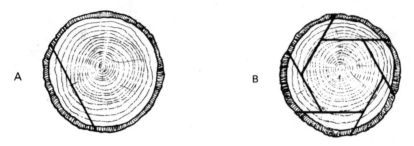

Fig. 4-6: Method of slab splitting a log.

One way of splitting heartwood of larger logs is the "daisy" technique. This method, as shown in Fig. 4-7A, is fast and does not usually require driving wedges. You simply work around the outside, splitting off the sections like daisy petals. Follow the basic marked cuts, but it is usually not necessary to move the log or walk around the chopping block.

Another popular method of splitting large logs is to cut them in pie-shaped pieces (Fig. 4-7B). With all splitting techniques, remember that the main purpose of doing the job is to expose as much wood as possible to air and to make the wood small enough to handle easily. Of course, the size of the pieces will depend on where they are to be used: fireplace, furnace, or stove. Incidentally, slab, pie, or daisy cut pieces of firewood can easily be cut into kindling. By the

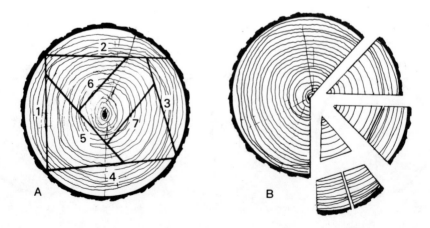

Fig. 4-7: Splitting a large log by (A) daisying and (B) cutting pie-shaped pieces.

Seasoning Firewood 79

Fig. 4-8: Proper way of cutting kindling with a hatchet.

way, it is usually safer to make the wood and blade of the hatchet (generally preferred for cutting kindling) descend together, as shown in Fig. 4-8.

As shown in Fig. 4-9, a crotched log is the best way to steady an uneven piece on the chopping block. If logs have branch stubs or knots in them, the split line should run through the center of the knots or stubs. Curves in the grain of the bark often indicate knots which have been overgrown and hidden. Looking for humps in the bark is another good way of spotting hidden knots. Some splitting experts advocate placing the piece to be cut on the block tree-top end up. Others prefer placing it base-end up. Use whatever way works best for you, but study and learn the natural grain of the wood and always cut with it.

While a crotched log may be helpful in supporting wobbly pieces on the block, splitting such a log can prove to be a problem. There are two ways to handle a crotched log. You can saw the legs apart (Fig. 4-10A) and split them individually (Fig. 4-10B), or stand the piece on its legs and drive a wedge in on line with the legs' centers (Fig. 4-10C).

Fig. 4-9: A crotched log will help steady an uneven log for splitting.

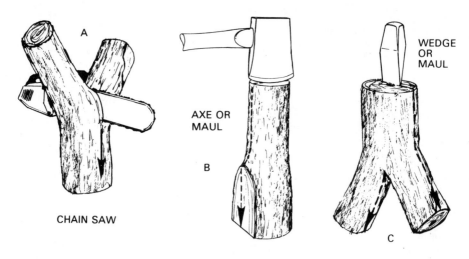

Fig. 4-10: Two methods of splitting a crotched log.

Some wood furnaces require long logs (36 to 48 inches in length). Figure 4-11A shows one method of splitting a log lengthwise using wedges. A chain saw can also be used to rip a log in half. Use a sawbuck (Fig. 4-11B) or stand the log on one end. But, regardless of the method used, be sure that the log is supported securely in place. When ripping vertically, keep your feet well back from the guide bar end projecting through the cut. It is often necessary to drive wedges in the saw kerf to prevent binding. Also, when ripping softwoods, you must remember that there is a tendency for the saw to pull long, stringy fibers from the wood. These fibers must occasionally be removed from the area around the sprocket and the clutch drum, as well as the chain.

Removing the bark from the logs will help speed along the seasoning of fuelwood, but this practice is too difficult and time-consuming to be practical. The exceptions to this are logs cut from an elm that died as a result of Dutch elm disease. In this case, be sure to remove all the bark immediately and burn it to kill the beetles which caused the disease. Contrary to popular belief, the bark contains about the same heat value as the rest of the wood.

Power Log Splitting

If the hand wood splitting method does not appeal to you, of course, you can always buy or rent a power splitting machine (see Chapter 2). They take much of the hard work out of splitting and do the job a great deal faster. The only work involved with most power splitters is placing the log on the splitter rack, moving away, and applying the power to start the splitting process. After the log is split, the split halves can be removed from the rack and the pieces stacked. Logs can also be quartered or cut in pie shapes with a power splitter.

When operating a power splitter, be sure to follow the manufacturer's instructions given in the owner's manual to the letter. To illustrate how a power splitter operates, let us take a look at how an electric powered hydraulic log splitter—very popular with weekend log cutters because of the convenient

Seasoning Firewood 81

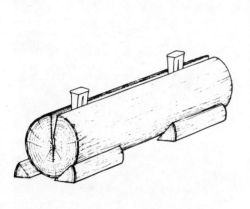

A   B

Fig. 4-11: Two methods of halving long logs.

power source—splits a log. Of course, before operating any power log splitter, it is very important to become familiar with all the controls. The controls of the electric powered hydraulic log splitter shown in Fig. 4-12A are as follows:

1. The *pressure release valve* is opened by turning the pressure release lever upwards (counterclockwise) with your foot. This will allow the ram guide to return to the down position. To split, the pressure release lever must be in the down position (closed).

2. The *wedge adjustment lever* must be depressed to adjust the wedge assembly. By releasing the lever, the wedge assembly will lock in place.

3. The *pedal switch,* when depressed onto the bellows, actuates the pressure switch. This starts the ram guide on its upward cycle. Release of the pedal switch will stop all movement of the ram guide. If ever a log becomes jammed on the wedge, and the ram guide stops its upward travel, remove your foot from the pedal switch. This will prevent any damage to the motor assembly.

4. The *ram guide* positions and pushes the log into the wedge.

To operate the illustrated electric powered hydraulic log splitter, proceed as follows:

1. Move the log splitter to a clear, level area (Fig. 4-12B) and set it in an upright position.

2. Connect the power cord and route it to the right (opposite the pedal switch side), away from the unit.

3. Place the log on the ram guide. Position the wedge assembly so that the point comes in contact with the center of the log (Fig. 4-12C). **Keep hands and fingers from between the log and the wedge.**

4. Stand to the left (pedal switch side) of the unit and position hands on each side of the log splitter about 15 to 20 inches away from log with the guide post between you and the log.

Fig. 4-12: The parts of an electric powered log splitter and how it splits a log.

5. Actuate the pedal switch, when the wedge enters the log about halfway; place one hand on each side of the log.

6. As the log splitter completes its split, guide the separated pieces safely to the ground (Fig. 4-12D). **Note:** Depending on the type and length of wood, the log splitter will go through its cycle without completely splitting the log. To complete the split, lower the ram guide by turning the pressure release lever counterclockwise. As the ram guide retracts, simultaneously lower the wedge, keeping the log in place. When the ram guide has fully retracted, release the wedge adjustment lever and reset the pressure release pedal. The unit is now ready to complete the split.

Odd shaped logs can cause problems with any power splitter unless they are correctly positioned. For example, when using the vertical log splitter illustrated in Fig. 4-12A, knots in logs should be placed closest to the ram guide, away from the wedge (Fig. 4-13A). The Y-shaped log should be placed only as shown in Fig. 4-13B. Never place the wedge in the center of the "Y" as this may damage your log splitter. Do not turn the log upside down. When splitting a hollow-centered

Seasoning Firewood 83

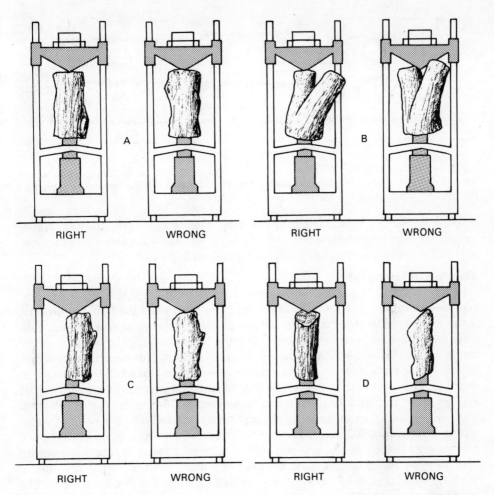

Fig. 4-13: Right and wrong way to split "problem" logs on a hydraulic log splitter.

log, be sure not to place the wedge or ram guide in the center of the hole. Instead, place the wedge and ram guide on the firm part of the log on either side of the hole (Fig. 4-13C). If one end of the log is not cut level, place the angle cut against the wedge and rest the most level end on the ram guide. If both ends of the log are not level, place the most level end on the bottom support in such a way that the angle of the cut is from the front to the back and not sideways (Fig. 4-13D).

Powered log splitters should be used **only** for splitting logs. Never use the power pack assembly (jack mechanism) for any other purpose. Also because of the basic design of log splitters, they should only be used to split logs lengthwise.

STACKING AND STORAGE OF FIREWOOD

After the firewood has been cut and split, it must be carefully stacked and stored to permit the wood to season properly. Of course, temperature, air humidity, and exposure to rain and snow all affect drying time. When drying wood, the greater the surface area exposed to the air, the more rapid the drying. To become

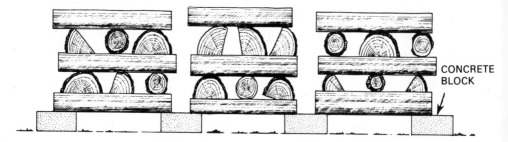

Fig. 4-14: Using concrete blocks as a base for stacking firewood. Concrete blocks never rot and prevent wicking. Eight-inch blocks are best as they keep the wood a bit higher off the ground, protecting the bottom layer.

properly seasoned, logs, after they are split, must be properly stacked. **Never throw firewood into a haphazard pile;** if you do, you will lose some logs to decay (rot) and will never know how much wood you really have.

Stacking of Firewood

Firewood should be placed in an open area to obtain rapid drying and to prevent deterioration. The best wood drying piles are raised well off the ground, either supported by lengths of log, old lumber, or concrete blocks. Cinder or concrete blocks are ideal since they eliminate the loss of any logs to ground rot and last for years (Fig. 4-14). Stack the logs loosely, allowing for maximum air circulation. The bottom layer should consist of halved logs placed slit side down on the support pieces. Stack the remaining halved, quartered, and smaller whole logs on top of this bed. Position these bark side up. (The curved water-resistant bark will act as a barrier against rain and help promote easy drainage of water through the woodpile.) Alternating each layer "log cabin" style (Fig. 4-15) will give support to the pile. Or, you can quickly construct a simple support system to hold your woodpile secure. End braces are a good solution to the problem of stacking wood (Fig. 4-16). Constructed with 2 by 4's or logs, end braces are like book ends and can be built to accurately measure a standard cord. The boards or logs beneath the woodpile keep the bottom row off of wet ground. Several commercial wood storage racks, such as the aluminum one shown in Fig. 4-17, are easy to put together.

Fig. 4-15: Alternate, or "log-cabin," stacking facilitates drying. The looser the stacking, the quicker the drying.

Seasoning Firewood 85

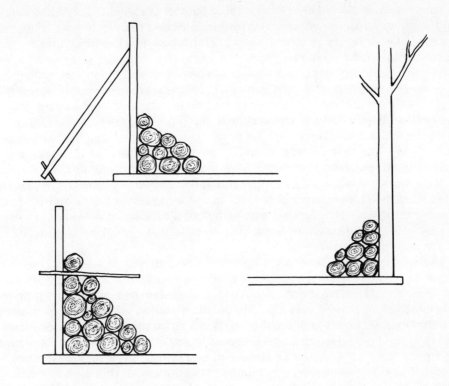

Fig. 4-16: Stacking wood on logs and using end braces of varying styles.

Fig. 4-17: A ready-to-assemble, aluminum wood storage rack.

Your woodpile should always be located outdoors facing the prevailing winds and in direct sunshine whenever possible. Stacking the wood directly against the side of your house or other building retards both air circulation and sun exposure time. Insect problems can also arise in this situation, so an open area, well away from any living quarters, is the best spot to season your wood.

Gentle rains will not hurt the seasoning process, but covering the top of the stack with a sheet of clear plastic will guard against damage from soaking downpours. Thin, 2-mil plastic is a bit too flimsy, but the 4-mil variety is sturdy enough to do the job and costs less than the 6-mil type. Remember that the black sheet absorbs the heat of the sun, creating a "greenhouse effect." That is, on sunny days, the temperature within the covering will rise considerably, facilitating quicker moisture evaporation. Proper ventilation and air circulation is necessary so that the moisture escapes and does not condense on the plastic and drip back down into the woodpile. Since any polyethylene plastic sheeting—either clear or black—will deteriorate when exposed to direct sunlight, you may have to replace your covering every few months.

Secure the polyethylene covering with bricks, stones, or logs. As shown in Fig. 4-18, roll the plastic sheet over a sapling-size pole which will hold one side down, or enable the plastic sheeting to be fastened to the logs. Be sure to orient the log stack in such a way that the plastic sheeting will face the prevailing winds. Oriented in this fashion, the winds will strike the sheeting, forcing it into the logs and holding it into place. Oriented in any other manner, the plastic will be ripped from its moorings by the wind, and rain or snow may pour onto the wood. There are, of course, any number of variations of this plan.

For example, solar wood driers expand this heating principle to an even greater extent. Complete seasoning is often possible in as little as three to four months. If you need your wood within a relatively short time, or the climate in your area is not the best for easy drying, a solar drier may solve some of your problems.

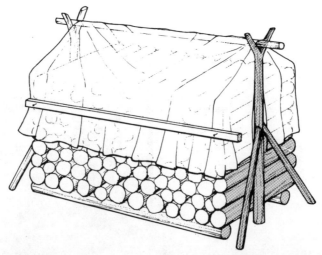

Fig. 4-18: Wood stack utilizing a polyethylene sheeting. The pole supports the sheeting, elevating it so that it does not touch the wood. This allows for free air circulation underneath the film. The pole may be permanently fixed or held up by a J-strap for easy removal. Be sure to brace the pole if you have a large stack.

Building a solar drier is not difficult. All it takes is some 4- or 6-mil polyethylene sheeting, a few pieces of sturdy framing lumber or logs, and a little time and common sense (Fig. 4-19). Again, the key is to keep the plastic sufficiently away from the wood to allow for proper air circulation. The high temperatures generated inside the dryer will hold insect problems to a minimum and keep the bark tight on the logs, resulting in a much cleaner fire. Heavy snow accumulations will rip the plastic, so a roof should be provided to keep the pile dry during the winter months.

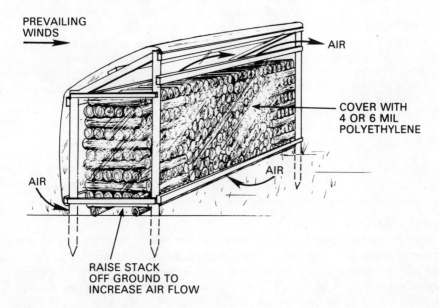

Fig. 4-19: A typical solar drier.

No matter which method you use, open air drying or solar heating, there are several ways that you can tell when your wood is properly seasoned and ready for burning or storage. As detailed in Table 4-3, cracks appearing in the ends of the logs are a good sign of well-seasoned wood. When struck together, dry logs will sound with a sharp, crisp ring; green wood produces a dull, muffled thud. Checking to see if firewood is still seasoning can be done by simply weighing a few pieces on a bathroom scale. Record the weight and return the pieces to the woodpile, marking them for easy identification. Reweigh these pieces a few weeks later. If there has been a significant weight loss, drying is still taking place and the wood still needs more time to mature.

Even though wood is properly stacked and covered for air or solar drying, it can become infested with several species of woodboring insects. These insects normally present no problem if the wood is burned within a year or two. However, one should be cautious about the amount of seasoned firewood that is stored in the home prior to burning. There is the possibility that powder post beetles and carpenter ants (Fig. 4-20) may be present in the wood and can migrate into the structural members of the house. Also watch out for other insects and spiders, such as the black widow species.

Table 4-3: GREEN AND DRY WOOD CHARACTERISTICS

	Green	Dry
Color: sapwood inner bark	Light; cream-colored. Light green to brown.	Shades of gray; dark. Dark brown to black.
Composition	Solid; slightly visible vascular cracks.	Definite and extensive vascular cracks.
Smell: hardwoods softwoods	"Woodsy"; pungent. "Piney."	Little or no odor. Slight "piney" odor, or no odor.
Sound (when struck with a stick)	Low-pitched thud; deadened sound.	High-pitched ring; cracking sound.
Touch	Wet; sticky sap; soft sap droplets.	Dry; no sticky residue; sap droplets are hard.
Bark	Bound to the wood.	Loose; separation areas.

Storing Firewood

Once the wood has properly seasoned, you may wish to move it to a more convenient storage area. Outside, wood will dry to between 14 and 25 percent moisture content depending on humidity, temperature, and wind. In a garage or woodshed it may dry to 10 to 15 percent moisture content; and wood may dry to between 5 and 12 percent in the house. However, the best places to store firewood are outdoors, under cover, and near the house. Storing wood inside the home may be the most convenient, but this practice has several serious drawbacks—insect infestation, dirt and dust problems, and an added fire hazard. Wood kept in basements often absorbs moisture rather than loses it and it could attract termites.

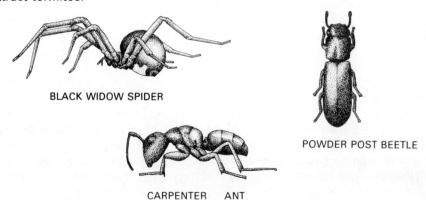

Fig. 4-20: Typical insects found in a wood pile.

Seasoning Firewood 89

Fig. 4-21: Several methods of storing wood.

A wood shed (Fig. 4-21), utility building, or free-standing garage is ideal. Firewood can also be stored in open areas, provided it is well covered with plastic or metal sheeting, or with a sturdy waterproof tarp, as previously described. As in seasoning, proper air circulation, protection from ground rot, and protection from severe weather are the most important factors.

90 Firewood and Your Chain Saw

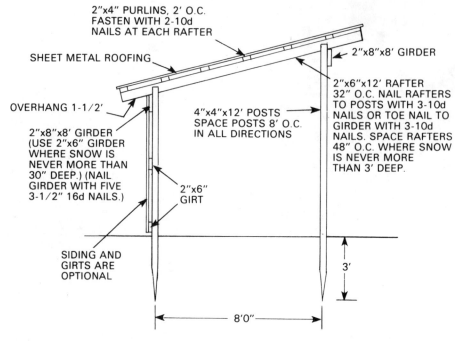

Fig. 4-22: Plans for a basic wood shed.

If sufficient space is available under a roof, seasoning and storage can be accomplished in one handling. This practice eliminates the extra handling of moving wood dried outside into a covered storage area. The wood shed plan illustrated in Fig. 4-22 is wide enough to provide for two rows of logs under the same roof. This permits you to maintain a wood burning rotation system—using the old wood first, while letting the fresher wood season. In fact, it is better to store your wood in at least two stacks or rows to provide a source of older wood. If you have a cord of wood in one stack, the oldest is usually on the bottom. To get

Fig. 4-23: Two methods of transporting logs into the house.

Seasoning Firewood 91

Fig. 4-24: Log tote you can make.

at it, you have to move practically the whole cord, which weighs at least a ton and a half, to get to the most seasoned wood. To say the very least, this is a lot of work—work that you do not need.

When bringing wood indoors, only take as much as you need for the day or evening. To make carrying the logs easier, you can use a log cart or dollie (Fig. 4-23), or you can make your own log totes, such as the ones shown in Fig. 4-24. They can be made either from 2-inch diameter logs or 1-inch thick by 1-5/8-inch wide rough lumber for a rustic look. Whichever you select, cut three lengths about 22 inches long. Then, drill four holes in each section; the diameter will depend on the thickness of the rope employed. (In the totes shown, 1/2-inch

A B

Fig. 4-25: Assembling log tote.

sisal rope was utilized and 3/4-inch holes drilled to provide ease of threading the rope through the wooded sections.) Drill one hole 2 inches away from each end, and two inboard holes equidistance apart as shown in Fig. 4-25A. In two of the log sections, drill two additional holes in the ends for attaching the handle.

After selecting a rope of desired thickness, cut four 5-1/2-foot lengths and two 4-foot handle pieces. Knot one end of each of the 5-1/2-foot lengths. Then, taking one wood section with a handle hole, thread one of the knotted ropes through the second hole from the end until the knot is snugly against it. Securely knot the rope inboard. Do the same with the three other knotted lengths, being sure to leave the handle free. Then, measure 8 inches from the inboard knots and loosely tie another row; these may need to be adjusted later. Add the middle wood piece and again loosely knot ropes on the outboard holes (Fig. 4-25B). Repeat this procedure for the other handle section; adjust all the knots until the tote is square, then tighten all of the ropes.

Knot an end of one of the 4-foot lengths, thread it up through one end hole and down through the opposite end hole of the same wood section, then knot it. Repeat the procedure for the other handle section. To prevent fraying, rope ends may be dipped in wax.

Once inside the house, wood buckets and baskets (Fig. 4-26) can be used to hold the logs. Metal containers, such as old coal buckets or antique copper clothes boilers, also make decorative and practical indoor wood holders. They are just the right size for an evening's worth of fireplace firewood.

Construction of an inside wood box with loading access from the outside woodpile is especially convenient if you plan to burn large amounts of wood. Be sure the access door from the outside of the house is tight-fitting to prevent heat loss. Always plan your fuel storage for a minimum of handling, but remember that storing large amounts of wood inside the home can lead to more problems than it can solve. A covered porch or patio is a much better place to store small amounts of seasoned wood. The log storage racks shown in Fig. 4-27 are ideal for such locations.

Fig. 4-26: Wood bucket and log holder.

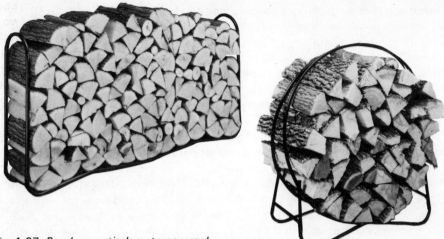

Fig. 4-27: Porch or patio log storage rack.

Cords Per Year

"How many cords of wood per year are used by the average person?" This is a difficult question to answer because of the number of variables: temperature zone, type of wood burner, species of wood burned, how well the house is insulated, frequency of wood burner use, severity of the winter, etc. In Table 4-4, it was assumed that seasoned (20 to 30 percent moisture), mixed hardwoods (oak, hickory, beech, maple, birch, etc.) were burned. If softwoods or inferior hardwoods are burned, the number of cords used will be increased 40 to 100 percent. The type of stove used in the cord test was an airtight stove, in good condition, with an efficiency rating of 50 percent. Stoves with lesser efficiencies, or stoves in less-than-fine condition, will increase the amount of wood used.

Table 4-4:
CORDS OF WOOD BURNED PER YEAR

Temperature Zone	Occasional, Light Use	Moderate, Intermittent Use	Heavy, Constant Use
Warm	1/2—1	2—3	3—5
Moderate	1—2	2—4	4—8
Cold	2—3	3—6	6—10

Another question frequently asked by weekend woodcutters is, "How much wood is there in a tree?" Again, there are many variables—size, species, growth, etc.—but most experts agree that a hardwood tree with a 12- to 14-inch base will generally yield half a cord of wood. Cut and split, the trunk will provide approximately a quarter cord; the branches will give another quarter.

THE YEAR-ROUND JOB OF THE WEEKEND WOODCUTTER

Cutting, hauling, splitting, and stacking firewood can be a year-round job. The cool autumn weather makes that time of the year the most enjoyable to work in, but do not plan on using any timber cut in the fall as firewood during the winter unless it is downed or dead and naturally well-seasoned. Felling trees in late spring and early summer has some advantages. Trees lose moisture through their leaves much faster than through their bark. So, if they are allowed to lay in place for a week or so before limbing, much of the moisture will escape through the leaves. Cutting in late spring and early summer also gives these trees the advantage of a few extra months of the hot sunshine necessary for proper seasoning. Whenever you decide to cut, remember that all fresh timber needs time to dry out before it is useful as fuelwood. In other words, it is best to plan and prepare ahead of time so that your firewood will be seasoned for at least 6 months before it is burned. But, remember that anytime is a good time for cutting and gathering your future fuelwood supplies. If you are a serious wood burner, woodcutting becomes as much a part of your life as burning it; the entire proposition is a 12-month task.

After reading this chapter, you may agree with us that the statement of Henry David Thoreau that wood heats twice was wrong. We believe that firewood warms you more than twice. It warms you:

- when you *fell* it,
- when you *buck* it,
- when you *haul* it,
- when you *cut* it to burning size,
- when you *split* it,
- when you *pile* it,
- when you *carry* it in, and
- when you *burn* it.

The latter, of course, is the best. Nothing beats the warm feeling of sitting and relaxing in front of the fire after the work is through, knowing that you are heating your home by your own efforts with the most intelligent fuel options available to humanity today.

Wood Stoves, Fireplaces, and Other Wood Burners 5

The characteristic of any wood burner appliance that usually receives the most attention is efficiency. Efficiency is the percentage or fraction of chemical energy available from the wood that heats the room. Efficiency depends on:
- the wood used
- the skill of the operator
- the design of the wood burner and chimney.

Wood varies in size, density, and moisture content; it is not a simple, uniform fuel like natural gas, propane, or fuel oil. Gas and oil burners uniformly mix fuel with oxygen, while variously sized chunks of wood are periodically dumped into a firebox. Over the years many stove designs have been developed to overcome the difficulties inherent in wood-fueled combustion.

The basics of heating with wood are relatively simple. In addition to firewood selection and procurement, stove or fireplace selection, safe installation practices, and proper operation procedures should be major considerations. First of all, select the wood burning appliance that best suits the needs of the household. Consider what portion of the house is to be heated. If the entire house is to be heated using existing hot air or hot water systems, consideration may be given to combination furnaces which will burn either wood or fossil fuels. If supplemental heating in one or two rooms is desired, then consider the multitude of fireplaces and accessories, radiant, and circulating wood burning stoves.

To help the consumer in selecting wood burning appliances, the Wood Heating Alliance (formerly called the Fireplace Institute) developed a voluntary efficiency testing program for wood stoves, fireplaces, and furnaces. The outcome of these tests at Auburn University resulted in the typical efficiency ranges for wood burning appliances that follow:

APPLIANCE	ESTIMATED EFFICIENCY RANGE
Masonry Fireplace	-10% to 10%
Manufactured Fireplace	-10% to 10%
Manufactured Fireplace With Circulation And Outside Combustion Air	10% to 45%
Free Standing Fireplace	-10% to 20%
Fireplace Stove	20% to 40%
Radiant Stove	50% to 70%
Circulator Stove	40% to 55%
Fireplace Insert	35% to 50%
Furnaces	40% to 60%

More on the Wood Heating Alliance tests can be found on page 118.

Gas and oil can be burned at fairly high efficiencies (generally between 65 and 85 percent) because the burner always operates at full output, but the fuel can easily be started and stopped. Wood-fueled heaters operate most efficiently when they are burning at nearly full capacity. In Spring and Fall, for instance, it is difficult to operate wood stoves at full output to create high enough temperatures for good combustion and heat transfer without overheating the room. Because the stove is normally operated at reduced draft in order to achieve a comfortable room temperature, or in order to hold the fire overnight, the efficiency of the stove is sacrificed. Thus, efficiency may not be the most important factor in selecting a wood burner. But, before taking a complete look at the various items that go into the selection of any wood burner, it is necessary to understand how wood burns. This knowledge will help in making your selection.

HOW WOOD BURNS

In any wood burning device, combustion transforms wood into heat, chemicals, and gases by chemical combination of hydrogen and carbon in the fuel with oxygen in the air. Complete combustion produces water vapor and carbon dioxide along with heat and noncombustible ashes. When incomplete combustion occurs, carbon monoxide, hydrocarbons, and other gases are formed.

As shown in Fig. 5-1, there are three stages or phases in the combustion of wood. Most wood stoves and other wood fuel appliances are designed to take advantage of these combustion phases.

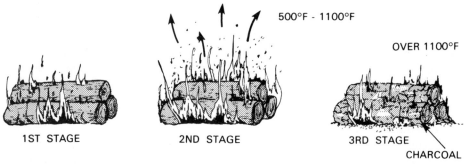

Fig. 5-1: The three stages of combustion.

In the first stage, moisture is evaporated from the wood and shows as white smoke (actually water vapor), with considerable heat being required to evaporate the water. The heat required for this stage does not warm the unit or the room. It is lost up the chimney.

In the second stage, the heated wood starts breaking down to form coals (charcoal). During this process, volatile wood components such as tars, acids, and combustible alcohols are driven off as gases, which in turn burn to form carbon dioxide and water. This stage requires considerable air (oxygen) and heat; the most efficient combustion occurs when air is directed around and over the burning fuel pile and a temperature of 1,100 degrees F is maintained in the combustion chamber. This air and temperature requirement has a considerable influence on stove design because it can determine stove efficiency to a large extent.

In the third stage, the burning coals maintain their required combustion temperature by means of heat conducted inward from the burning wood surface and heat radiated from adjoining pieces of burning wood. It is in this stage that the combustion cycle is chiefly maintained and useful home heat is generated.

All three stages of burning may occur at the same time. However, the first two usually occur when the fire is started or when wood is added. For efficient burning, the volatiles must be mixed with air and kept at a high temperature to burn completely inside the heating unit. Wood burns with a long, yellow flame, as opposed to hard coal which burns with a short blue flame. Many wood stoves do not make provision for this long flame and these gases escape up the chimney where their heating value is lost. By providing a long flame path, the heat from these burning gases can be utilized.

To obtain the efficient burning mentioned above, the fire needs combustion air or oxygen at two levels. First, it requires primary air entering either at the wood's own level or from beneath it (Fig. 5-2). This primary air mixes with the gases and moisture as they leave the wood during the first stage of combustion.

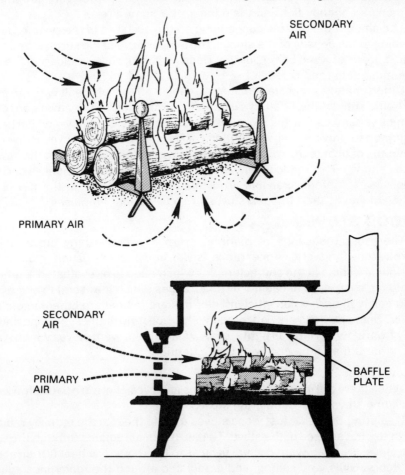

Fig. 5-2: The way primary and secondary air function in a fireplace and in an airtight wood stove.

The primary air continues to furnish oxygen for the other two stages; but the secondary air entering above the logs plays the important role in stages two and three of igniting the unburned volatiles borne by the updraft. If this secondary air does not mix before the unburned volatile gases drop below their ignition temperatures, no heat will result; the unburned volatiles will continue to travel up the chimney without producing heat. In other words, the more volatiles wasted up the chimney, the lower the efficiency of the wood burning appliance.

Creosote. No discussion of how wood burns would be complete without mention of the creosote problem. When wood burns, the combustion process is never absolutely finished. The smoke usually contains a substance called creosote which is dark brown or black and has an unpleasant odor. Its chemical composition is not well known because it is a very complex mixture of compounds.

When the stove pipe or chimney flue temperature drops below 250 degrees F, creosote will condense on the surfaces. At very low temperatures, below 150 degrees F, the creosote deposit is quite fluid. As these deposits are warmed, they coagulate and form a sticky tar-like substance which, when very hot, will ignite, causing a chimney fire and the danger of a home fire.

The amount of creosote condensing on the surfaces of the system varies according to the density of the smoke, the temperature of the surface, and the type and dryness of wood being burned. Dense smoke from a smoldering fire carries the most unburned creosote.

Unfortunately, creosote problems are most effectively dealt with by reducing the efficiency of the heating system. Air circulating in the stove causes more complete combustion and more heat escaping up the chimney—which heats the chimney to prevent creosote buildup. The more efficient stoves deliver larger amounts of heat to the room, therefore reducing temperatures in the stove pipe and chimney. This reduced temperature also increases the chances of creosote deposits. Therefore, creosote problems are more severe in the newer, more efficient stoves than in open stoves or conventional fireplaces.

WOOD STOVES

There are thousands of manufacturers of woodburning stoves (Fig. 5-3) throughout the world. Appearance, style, finish, construction, materials, and weight are some of the characteristics which need to be evaluated. Durability of welding, sharpness of cabinet or stove edges which may scratch or cut people, and ability to burn wood efficiently for maximum heat are other factors to consider. Some persons are concerned with creosoting of the stove pipe and chimney from slow burning airtight stoves, while others want stoves which will burn wood for a long time and with little attention.

Stove Types

While there are many different designs of stoves, there are basically two types: radiating stoves and circulating stoves.

Radiating Stoves. Most wood stoves transfer heat to the room by radiating it from the hot surface of the stove. That is, the heat radiates through the walls of the stove directly toward cooler surfaces in the room. The heat is then absorbed by floors, walls, and ceiling, and is radiated around the room once again.

Radiant heaters (Fig. 5-4) produce heat that is most intense at close proximity and diminishes rapidly with distance from the stove. Surfaces in direct line with

Fig. 5-3: Several styles of wood stoves.

Fig. 5-4: Typical radiant wood stove.

the stove will be heated. The heat is then absorbed by objects in the room, rather than warming the air directly.

Circulating Stoves. These stoves, sometimes called convection stoves, are constructed with a metal box spaced about 1 inch from the wall of the firebox (Fig. 5-5). Vents in the top and bottom of the outer box allow natural air currents to carry the heat away from the stove. The outer surface of a circulating stove is not as hot as a radiant stove and thus can be installed closer to combustible material than radiant stoves. Circulating stoves are better suited to heating a large room than radiant stoves.

A few stoves combine both the radiant and circulatory heat principles. Such stoves contain a blower system which constantly circulates air along the bottom of the stove, through heat exchanger tubes, and then out into the room.

For good heat-transfer efficiency with either type, certain design features should be investigated before purchasing a wood burning stove:

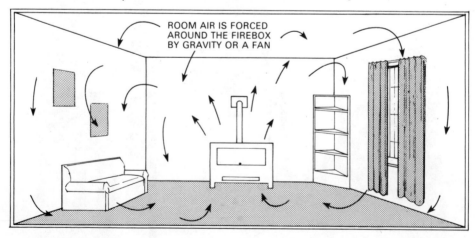

Fig. 5-5: Typical circulating wood stove.

1. The stove should have a large surface area to collect the heat and radiate it into the room.
2. The stove should have an airtight firebox with an adjustable draft control so that wood combustion can be closely controlled.
3. The stove should have internal baffles that aid in creating high temperature and turbulence in order for the stove to achieve maximum heat transfer to the radiating surfaces.
4. The heated gases should be contained in the stove as long as possible (by means of heat or by means of smoke chambers and baffles) in order to effect maximum heat transfer to the radiating surfaces.

In general, a large radiating surface area relative to the size of the fire is important in obtaining optimum heat-transfer efficiency.

Materials of Construction

When selecting a stove, you have a choice of three materials:

1. *Cast iron* stoves are made by pouring molten pig iron into a mold forming the components of a stove. These pieces are machined to fit, sandblasted and acid dripped (pickled) to remove scale and corrosion, assembled, then either enameled or blackened.
2. *Plate steel* stoves are heavy sheets of steel (typically 1/4- to 1/2-inch thick) cut to size, then welded together. They are wire brushed or pickled, then blackened.
3. *Sheet metal* is thin sheet steel (less than 1/8-inch thick) that is either welded or bolted to form the stove.

The material characteristics between plate steel and cast iron are not different enough to warrant significant concern. The thicker the metal, the longer the stove should last. Both materials retain roughly the same amount of heat per pound, and both stove types cost roughly the same. One difference is that a cast iron stove can be dressed up with designs in the casting.

A well-made stove will have clean castings, smooth welds, and good workmanship. Some stoves have firebricks or metal plates to prevent burnout; this may increase the lifetime of the stove and also increases the thermal mass (the storage medium for the heat—a 500-pound stove can continue to give off usable heat 4 hours after the fire is out). Cast and plate iron, while resistant to warping, are brittle and can be cracked easily by a hard blow, as, for instance, if a heavy object is dropped on either. Cast iron stoves are seldom airtight. If you wish your cast iron stove to be airtight, you must make sure that all joints are sealed with stove putty. Stove putty has a short life span; it must constantly be replaced with new putty.

Sheet metal stoves have been used for many years for heating. They are relatively inexpensive but have a shorter life than plate steel or cast iron stoves. They will quickly heat a room, but they also cool rapidly when the fire dies down. If occasional quick heating is needed, such as for a cabin, a garage, or emergency use, thin walled stoves are appropriate.

Stoves made of sheet steel can be made airtight, or virtually so, thus increasing the efficiency of the unit. Sheet steel will not crack, as will cast iron, but is subject to warpage at high temperatures. Sheet steel also gets extremely hot, and severe burns are much more likely than with cast iron. Sheet steel stoves are usually not as aesthetically pleasing as cast iron units. They tend to have a boxy appearance in order to accommodate tight seams.

To protect the walls of sheet metal stoves, they are sometimes lined with firebrick. This lining also permits a hotter fire in the firebox, but it reduces the radiation from that part of the stove.

In general, the material a stove is made from is secondary to workmanship, draft control, and design. One criterion for good workmanship is how the sections of the stove are joined together. Sheet steel stoves, for instance, should have continuous welds, not just spot welds. Cast or plate iron stoves should have tightly fitting joints that will help hold the joint compound in place.

Another good indication of quality in any wood burning stove—whether it be the circulating or radiating type—is its weight or mass. Remember, however, that the circulating type—because they dissipate heat faster—can be made of lighter-gauge material than the heavier and sturdier radiant counterparts.

Draft Control

A comfortable room temperature is most easily maintained by controlling the rate at which the fire consumes the wood in the stove. The loose fitting doors on a Franklin stove provide some control of the fire. However, much better control is achieved if the stove door seals the opening and the air flow into the stove is controlled by an adjustable vent. Stoves constructed in this manner are called airtight stoves.

The primary air supply may not be adequate or in the correct position to supply air to support the combustion of the volatile gases, so some high-efficiency stoves have an additional air vent to introduce secondary air above the flame. The amount of secondary air admitted is critical for efficient operation. If too little air is admitted, incomplete combustion will result. If too much air is admitted, the gases will be cooled, affecting combustion and heat transfer.

As shown in Fig. 5-6, there are six basic draft configurations. While manufacturers may vary these configurations slightly in an effort to make their models more efficient, the draft conditions described here are common for the way air moves through the stove.

In updraft stoves (Fig. 5-6A), the air enters through inlets at the bottom of the stove, then moves up through the burning fire and out of the flue. Most antique-styled stoves, particularly the potbellies, operate on this draft principle.

Diagonal stoves (Fig. 5-6B) have the air entering the bottom of the stove and then moving diagonally through the wood to the flue in the back of the stove. This pattern is typical of most drum (barrel) and box stoves.

In cross-draft stoves (Fig. 5-6C), the primary air enters the firebox area near the bottom and exits near the bottom at the back of the stove. The baffle which is built into the firebox, when heated by the fire, creates a long flame path which helps to efficiently burn the volatile gases that produce heat. Cross-draft stoves also require a bypass damper near the top of the smoke chamber to keep smoke from entering the room when the door is opened. There is a direct route from the top of the fuel chamber to the chimney when this damper is open. As more air is supplied to the fire, the smoke tends to be sucked out of the chamber instead of blowing into the room.

The down-draft stove (Fig. 5-6D) brings the entering primary air through the logs and forces it out at the bottom of the back of the stove. Most down-

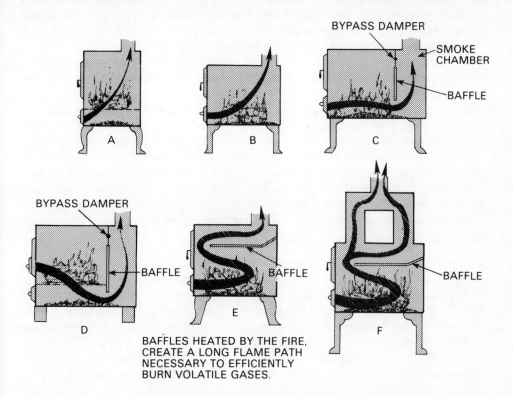

Fig. 5-6: Typical draft configurations.

draft stoves work more efficiently if an electric fan is used to pull the entering air down through the logs. Many airtight style stoves use either the cross-draft or down-draft air control system.

In S-draft stoves (Fig. 5-6E), air enters in the front of the stove and passes over the burning logs. Then, the unburned gases pass back over the flames to exit in front of the baffle (or smoke shelf). When the unburned gases and entering air come in contact, a high-efficiency burn is achieved when the stove is operating at high temperatures.

The double chamber stove (Fig. 5-6F) is a variation of the S-draft system that provides a separate smoke chamber rather than a mere baffle plate. The S-draft and double chamber arrangements are usually employed with most high-efficiency stoves, including the so-called Scandinavian models.

Manual or Automatic Draft Controls. The draft controls can be operated manually or automatically. With manual controls, you have to open and close doors and air intake regulators, manipulate dampers, and generally fuss with the stove to keep it going. As a result, the fire in a manually controlled stove can be held for long burns. This is helped along still further by the fact that most manual draft-control stoves have no grates. The lack of grates permits base-burning, which helps to even out or decrease the burn rate, since no incoming air gets to the fire from below.

For most automatic draft controls, bimetal thermostats are employed. This bimetal device relaxes as it is heated and contracts as it cools. In this way it can hold a preset level of heat output and thereby control the rate of burn. As the stove warms up, the bimetal control closes off the air supply; as it cools down, it allows more air in to keep it burning at the desired level.

Unlike automatic thermostatic controls used with oil or gas furnaces, you cannot expect to start with a cold stove, load it up, light it, and immediately set the thermostat to "medium." The stove must be manually brought up to the desired operating temperature before the automatic thermostat is going to function properly. Also, they have the disadvantage that when the fire goes out, the draft is usually wide open. Since thermostatically controlled stoves are generally designed to be used with grates, this almost always means that the fire will go completely out if not fueled in time. Also, the thermostat is probably the most vulnerable part on most automatically controlled stoves. It frequently gets stuck in either the closed or open position. If stuck in the closed position, it simply means no fire. If stuck in the open position, however, the unit could be overheated.

Stove Heating Capacity

Many manufacturers rate their stoves by either the number of rooms the stove will heat or the number of cubic feet. In the latter case, the simple rule of thumb is as follows: Multiply the length, width, and height of the area you wish to heat by 5.5. The result will roughly represent the Btu's/hour output required to heat the area. If, for example, you wish to heat a 15 by 20 foot family room with 8-foot ceilings, multiply 15 by 20 by 8 by 5.5. The result is 13,200 Btu's. Thus, you should select a stove that delivers between 13,000 and 18,000 Btu's/hour. This should give you sufficient capacity to heat the immediate area as well as provide some heat for adjoining spaces.

Keep in mind, however, that any capacity rating must be used cautiously. In colder areas or poorly insulated houses, less area will be heated. In addition, heat movement into more than one room of the house may be very poor. Unless the rooms are very open, attempting to heat more than one room with a stove may result in uneven temperatures and cold drafts along the floor. In cases such as these, many heating experts recommend that the Btu burning rate be approximately one third to one half of that which you can expect from good stove performance over the long run. Figuring size this way gives you reserve capacity for the coldest days, but allows you to burn at a rate high enough to provide heat for the space you want heated.

This advice by the experts must be taken with a little warning. All too many people purchase a stove that is too large for the room in which it is installed. If the stove is operating efficiently, the room is overheated. It is difficult to maintain a small fire in a large stove. One option would be to try to move the heat into other rooms of the house.

While the concepts of heat flow are well established, the design recommendations for moving heat from a wood stove through the house have not been well documented. Warm air will rise to a room above if a path is provided for the cool air to return to the stove (Fig. 5-7). The return path could be registers close to the outside walls, or an open stairway.

As stated earlier, some stoves come equipped with a blower system that will provide forced convection. While such a stove heats in the same manner as a conventional hot-air furnace, it has one major difference. The normal hot-air

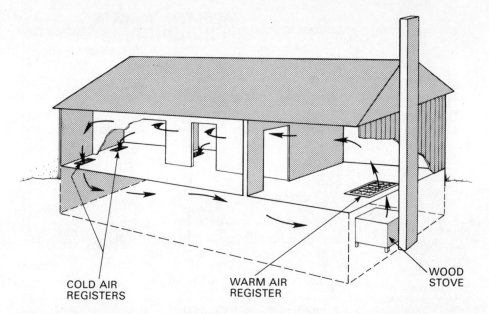

Fig. 5-7: Natural convection system using a basement wood stove. The size of both the hot air and cold air return registers can be smaller than the registers used for natural convection.

furnace fan is designed to move large amounts of hot air as the furnace is running intermittently. On the other hand, a wood stove with a blower is a gentle, continuous heat source.

If your stove is not equipped with a blower system, it is frequently possible to add a fan to your unit. Typical specifications for such a fan should usually be a centrifugal (squirrel cage) fan with a fractional horsepower motor for continuous duty. The fan should circulate about one complete air change in the house each 30 minutes. A 300 to 500 cfm fan would be adequate for small homes. For more details on adding a blower system to your stove, check your local wood stove dealer.

Another effective way to move heat to other parts of the house is to install hot water coils in the stove and pump the hot water to baseboard radiators. The water coils will normally be kept at a temperature that is less than the boiling point of water. Creosote will form on these coils but creosote is less of a problem on coils inside the stove than coils inside the stove pipe. A pressure relief valve should always be installed in the water line. Such valves are part of all conventional water heaters and boilers.

With some installations, the hot water may flow through the pipes by natural convection. However, the pipes must be carefully designed so that air traps are avoided. A thermostatically controlled pump provides positive circulation and may be more effective than natural convection systems (Fig. 5-8).

Domestic water heating is also possible using your wood burning stove. A typical domestic hot water system, using a built-in heat exchanger, is shown in Fig. 5-9A. Stovepipe heat exchangers (Fig. 5-9B) are also available to be used when it is not possible to install a heat exchanger in the stove's firebox because

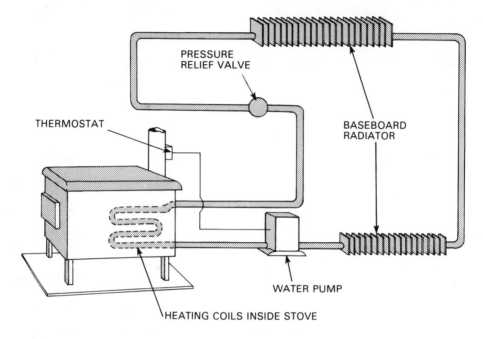

Fig. 5-8: Typical hot-water heating system employing a stove.

of the shape of the stove. Both types of heat exchangers can be used as a backup system for conventional hot water heaters powered by gas, electrical, or solar energy (Fig. 5-9C). A fireplace (Fig. 5-9D) with a built-in heat exchanger is also a source of domestic hot water.

The operation of all heat exchanger systems is the same. The cold water from the water supply enters the storage tank at the bottom (w). This cold water leaves the tank at "x," travels upward through the heat exchanger coil, and reenters the storage tank at "y." This cycle of taking cold water from the bottom of the tank and discharging hot water at the top of the tank will continue until the storage tank is filled with hot water. This hot water can be drawn from the tank at "z" to furnish it to the fixtures in the plumbing system. A 30- to 40-gallon storage tank is adequate for a family of four. If properly insulated, the water will remain hot for 36 to 48 hours.

Outside Air Inlets. A further complication to efficient stove operation has occurred in tightly constructed, well insulated houses. These houses may not have enough air leakage to supply air to the wood stove, causing poor burning and smoke. Opening a nearby window normally solves the problem, but this lets in cold air that, in effect, lowers heating efficiency.

Other Factors

When purchasing a wood stove, there are several design and quality factors to keep in mind. For example, the way the stove is loaded is an important consideration, mainly because of the smoke problem. That is, while top-loading stoves are easier to load and fill, they are more likely to let smoke into the room. For this reason, most stoves are loaded from the front, usually near the bottom. While such a location makes it harder to insert large logs and to fill the firebox completely, it greatly reduces the amount of smoke that can escape into the room. Also, the size of the doors is another factor in the smoke problem. While large doors make adding wood to the fire much easier, there is a greater likelihood of smoke escaping when they are opened.

Another feature to consider is that of ash removal. Most species of wood, when burned in airtight stoves, yield approximately 1 to 4 percent of their original volume in ash. Thus, most wood stoves will require the ash to be removed every 4 to 10 days. Ash residue is extremely beneficial as a fertilizer and should be utilized as such whenever possible. Ashes are used in gardens, flower beds, flower boxes, lawns, or icy sidewalks.

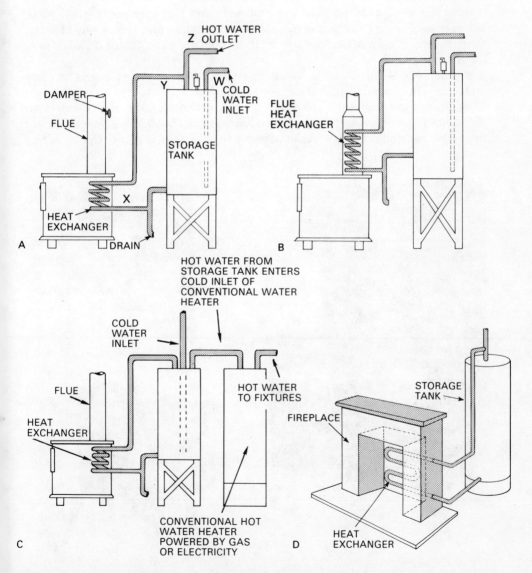

Fig. 5-9: Typical domestic hot-water heating systems.

Ash removal is easier on stoves equipped with grates. (The grates should be cast iron.) Usually, grated stoves have a pan arrangement which can be removed from the stove Be sure to dispose of ash in a noncombustible container or in an outside dumping area. Since non-grated or base-burning stoves have no room for an ash pan, the ashes must be cleaned out with a shovel.

When selecting a stove, the following considerations are rated in priority order:
1. Decide on what its end use will be.
2. Decide on the location of the stove.
3. Select the correct size for your application.
4. Pick a general type that best fits your use.
5. Talk to stove dealers; tell them what you want and why.
6. Pick a stove that fits your use and taste, not just your pocketbook. Also, select the stove on the basis of quality of construction and availability of parts. Consider the stove a major investment that could pay a handsome dividend over its lifetime.

Kitchen (Cooking) Stoves. No commentary on wood stoves would be complete without some mention of kitchen stoves. Perhaps no other appliance conjures up so many romantic images of rugged, yet comfortable, early American home life than these huge, rangy wood stoves (Fig. 5-10). This image is more than a bit misleading. Since it is impossible to cook anything until the stove has

Fig. 5-10: Typical kitchen stoves.

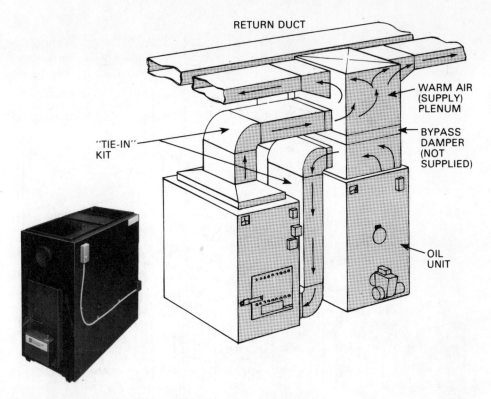

Fig. 5-11: Typical warm-air wood burning furnace system.

reached minimum cooking temperatures (about 250 degrees F), it could be an hour before you can so much as fry an egg. The stove burns an inordinate amount of wood and needs a lot of attention. In addition, it takes a great deal of skill and determination to make a complete meal. The oven section is heated on one side only, so you must wait until the entire compartment has reached the desired temperature before you can put in your cake or pie.

Kitchen stoves are not airtight and are only about 30 to 35 percent efficient. Most cooking stoves or ranges are of cast iron construction and will burn either wood or coal. When wood is burned, the pieces need to be short and split fairly small.

WOOD FURNACES

Wood stoves are good for heating small, open houses or one or two rooms. But most houses are designed for central heating with a furnace. Wood burning furnaces are large enough to heat an entire house. There are types available that will heat any size home and are designed for using either hot air or water transfer mediums. In most cases, the furnace is designed to replace the existing furnace, using the pipes or vents currently in place. Also, some units are designed to provide supplemental heat with an existing oil or gas central system (Fig. 5-11 and 5-12). In other words, a good wood furnace with a conventional heat circulating system provides even heat distribution throughout the house and offers the convenience of remote temperature control.

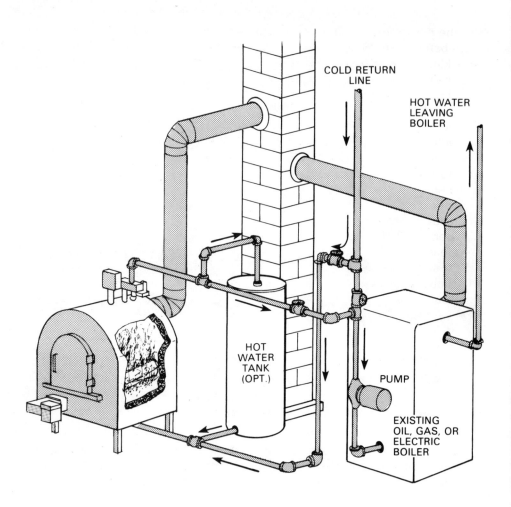

Fig. 5-12: Typical hookup of an oil/gas backup system.

Because of the large fuel capacity and controlled burning, most furnaces require stoking no more often than once every 10 hours and sometimes as little as once every 24 hours. Some of these units have a storage magazine like a hopper, which holds a large charge of wood and feeds it slowly into the combustion zone. Others have extremely large fireboxes that permit the stoking of large quantities of large chunks of wood (as large as 13 inches in diameter and 3 feet long). Though burning large chunks is not as efficient as burning small sticks, the saving in cost of fuel preparation is great.

All furnaces are thermostatically controlled. When the thermostat is not calling for heat, the primary air supply is very low, so the fire will continue with a very low heat output. When heat is demanded by the thermostat, the primary and secondary air supplies are opened. The fire will burn at a high rate, with near complete combustion. On some furnaces, primary and secondary air is supplied by a blower. The hot gases of combustion then pass through a heat exchanger to heat the air or water used to heat the house. Usually another blower forces air from the house through the exchanger. This type of furnace can be quite efficient—better than 70 percent.

In addition to cutting the wood and stoking the fire, a wood furnace requires more maintenance than a gas or oil burning one. Ashes must be removed and the chimney must be cleaned periodically. The boiler or heat exchanger may need cleaning to remove any deposits. Before purchasing a wood burning furnace, you should first talk to someone who already has one to learn more about how well they operate.

Most people want to keep their house above freezing when they are not at home to keep the fire burning. Many wood burning furnaces come equipped, or can be equipped, with thermostatically controlled oil burners. The oil furnace automatically comes on when the wood fire dies down.

Installation of either a wood furnace operated by itself or with a gas/oil back-up system should be done as directed by the manufacturer. Also, take time to plan the installation of your wood burning furnace or boiler so that it conforms with the safety requirements of local building codes. Chimneys must be of sound construction and, along with the smoke pipe from the heater, must be cleaned periodically. This will remove creosote deposits that choke the draft and create a chimney fire hazard.

Multi-fuel Furnaces. While single multi-fuel furnaces (wood/oil or wood/gas) have been popular in Europe for years, they have been used only recently in the United States (Fig. 5-13). Multi-fuel furnaces can operate solely on oil or gas, and they can augment the wood whenever it is not giving off sufficient heat to satisfy the thermostat. Some units have separate combustion chambers, while others are designed to be fired directly into wood coals without fear of explosion.

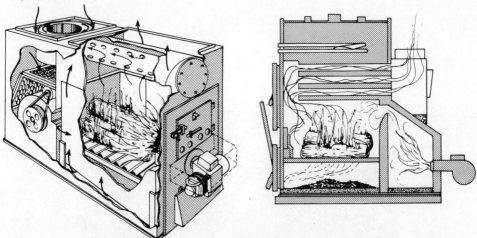

Fig. 5-13: Operation of a typical multi-fuel unit.

Most multi-fuel furnaces are only slightly more expensive than comparable single-fuel furnaces. But, before considering multi-fuel furnaces too seriously, check local building or fire codes. In some areas, they may not be legal since code requirements have not been suitably developed to ensure safe construction and maintenance.

FIREPLACES

A masonry fireplace supplies radiant energy to bring quick comfort to a cold room. However, it is not a very efficient heater. The volatile gases, containing the highest percentage of the heat value of the wood, are drawn up the chimney unburned and wasted to the outside air. In addition, the draft created by the fire draws room air up the chimney along with the burning gases, sometimes resulting in a net heat loss for the whole house. Additional heat is lost if the damper is left open after the fire dies out. As mentioned earlier in this chapter, a typical masonry or metal manufactured fireplace has an efficiency of 10 percent or less.

Both masonry and "zero clearance" prefabricated metal fireplace units can be made more efficient by considering any or all of the following improvements:

1. Sliding glass fronts that allow you to view the fire and still have the unit operating at top efficiency.

2. Air-circulating devices, usually in the form of tubes which take in cool air at the bottom and shoot hot air out at the top. These air-circulators may be either passive—working solely from the force of the draft—or active—relying upon electric powered fans to pull in the cool air and shoot out the hot air.

3. Water pipes passing through the unit to provide hot water for heating purposes or provide a supply of hot water.

4. An intake opening at the bottom, usually through the ashpit, to draw cool air into the unit from outside the house so that already warmed room air will not be used to provide oxygen for the fire.

In recent years, some of the better metal zero clearance prefabricated fireplace units (Fig. 5-14) featured air/heat circulation systems that incorporate outside combustion air. Such a system greatly increases the efficiency of the

Fig. 5-14: Typical "zero clearance" prefabricated metal fire unit. Such fireplace units can go right up against the wall, but make sure they bear the Underwriter's Laboratories, Inc. (UL) label that states they are listed for zero clearance installation.

fireplace. Zero clearance, of course, means that the fireplace may be installed on or against existing walls and floors. The exterior of the heavy metal insulated fire chamber remains safely cool while the fire burns. They can be installed by the average do-it-yourselfer in a few hours with a minimum of tools. Zero clearance units will usually enhance the appearance of the room, especially those with sliding glass doors. They can be supported and surrounded with standard building materials, combustible or not. They can be installed without any expensive masonry work or any of the mess that masonry fireplace construction entails. But, of course, you can trim your newly installed unit with masonry or any other finishing material—wood paneling, stucco, or shingles (Fig. 5-15).

Fig. 5-15: Finished zero clearance fireplace installations.

With most modern zero clearance fireplaces, cool air from the floor and/or from the outside of the house is taken in through vents in the front or side of the fireplace and channeled through heat exchange chambers. The fire in the firebox heats the air and, by gravity feed, forces it up and out through air vents or outlets in the top of the fireplace (Fig. 5-16); it then goes back into the same or adjoining room. To increase the flow of air through the heat exchange chambers, a fan is sometimes placed in the duct system.

Heat circulating zero clearance units are not as efficient as airtight stoves, usually giving up to 45 percent efficiency. The efficiency of a standard fireplace can be increased by installing an insert. Such an insert will permit you to use your fireplace at any time, even when the temperature really dips. Without an insert, at temperatures below 20 degrees F, your ordinary fireplace is actually cooling your house rather than heating it, even while burning wood at a furious rate. Since there are so many zero clearance units on the market, each with its own installation instructions, the best advice is to read and carefully follow all manufacturer's instructions to the letter. Following instructions exactly is necessary to validate your fire insurance, comply with local fire codes, and to insure the efficient operation of the unit.

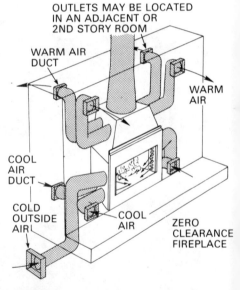

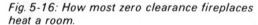
Fig. 5-16: How most zero clearance fireplaces heat a room.

Metal free-standing fireplaces (Fig. 5-17) have also proven popular for a number of reasons, not the least of which is the appeal of open fires. With the free-standing unit, many people, instead of just a few, can enjoy the warmth of the flames. Free-standing units can be installed in homes which have not been equipped with a built-in masonry fireplace. Generally, free-standing units are inexpensive and can be installed easily by the do-it-yourselfer. Basically they are used for casual or secondary heating purposes. Free-standing fireplaces should not be installed as basic home heaters unless they can be sealed tight, with controlled air inlets. These metal fireplaces have a little better efficiency than masonry fireplaces. They warm up faster, but they cool rapidly once the fire dies down.

Fig. 5-17: Various styles of metal free-standing fireplaces.

Modifications to Existing Fireplaces

Modifications to a fireplace can include the installation of glass doors (Fig. 5-18). While tight glass doors on the fireplace will greatly reduce the radiation that reaches the room, the doors reduce the amount of warm air from the room that is lost up the chimney. Probably the most heat will be gained if the doors are opened during the hotter stages of the fire and closed as the fire dies out. The greatest saving occurs when the closed doors control the loss of heat overnight. A tight fitting metal door or sheet of asbestos could also be used.

Fan heat exchangers are available that help to increase the efficiency of your fireplace. As shown in Fig. 5-19, the twin squirrel-cage blowers draw room air through the heat exchangers of the fireplace heater and gently circulate warmed air throughout the room. The heat exchangers are sealed, preventing flue gases from being circulated, and dampers in each exchanger prevent convection of warm air when the unit is shut off.

ig. 5-18: Tight fitting glass doors such as shown here reduce the amount of warm air from the room that is lost up the chimney.

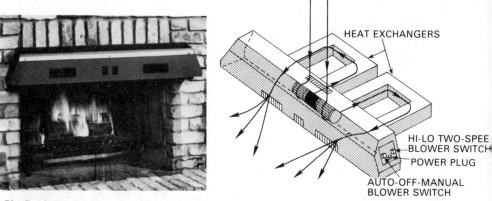

Fig. 5-19: Typical fan heat exchanger and how it works.

Hollow grate or tube heat exchangers (Fig. 5-20) can also be had that can be placed in the fireplace to provide additional heat by air circulation. That is, a series of C-shaped hollow tubes are used in place of a conventional fireplace grate. As the fire warms up, the air inside the tubes is heated from the room temperature to 600 degrees F or more. Cool air flows in the bottom of the grate and passes out of the top. Some units use a small fan to pull the cool air into the bottom and force it out of the top. By adding a hollow tube grate heat exchanger or fan heat exchanger along with glass doors, you can improve the efficiency of your present fireplace up to 30 percent.

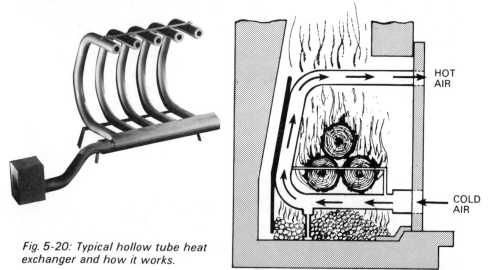

Fig. 5-20: Typical hollow tube heat exchanger and how it works.

Firebrick makes a perfect lining for a fireplace if all you are interested in is a solid, fireproof firebox. But, firebrick absorbs heat, radiates little, and reflects none. So your heat either stays in the firebox or goes up the chimney. One way to reflect more of the heat into the room is to place a curved, sheet metal reflector (Fig. 5-21) at the back of your fireplace. The reflector not only reflects more heat, but improves the draft at the same time by directing the gases toward the damper. Efficiency does not match that of heat exchangers, but even 10 percent would double the efficiency of the average fireplace.

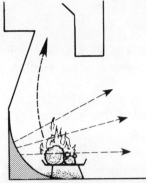

Fig. 5-21: Typical fireplace reflector.

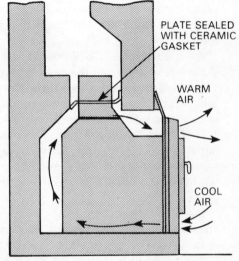

Fig. 5-22: Typical stove inserted in a fireplace and how it works.

The atmosphere of an open fire is by far the most important reason people buy or install fireplaces. The fireplace is not an efficient heating device even with extensive modifications. Except for supplemental or emergency usage, the fireplace is best left to being a pleasant experience.

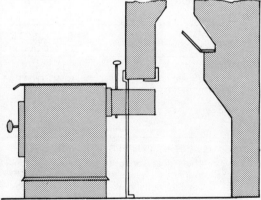

Fig. 5-23: Installation of a typical in-front fireplace insert stove.

Replacing Fireplaces with Wood Stoves

Several stoves have been developed to increase the efficiency of the fireplace. Some controlled draft wood stoves are built to be inserted inside existing fireplaces (Fig. 5-22). Others are designed to sit in front of the fireplace (Fig. 5-23) with a piece of sheet metal to cover the fireplace opening. These installations, as detailed in the next chapter, eliminate the need for another chimney.

WOOD BURNING APPLIANCE RATINGS

Manufacturers of fireplace units, stoves, fireplace inserts, and other wood burning appliances, who submit units to the Auburn University lab for testing (see page 95) and complete the license agreement, are then allowed to display an efficiency rating label for the consumer's benefit. The information of importance to a would-be purchaser that appears on an efficiency rating label is as follows:

A—Manufacturer's Name, Model Number. The manufacturer's name and the model number of the tested unit will be printed here.

B—Test Number. Three firing rates are tested in an attempt to match the intensities of wood burning that consumers might use in their homes. Number 1 refers to low fire; number 2 to medium fire; number 3 to high fire.

C—Wood Pounds/Hour. This column tells you the pounds of wood burned per hour during testing. These figures will tell you how much wood must be burned to achieve the heat output needed for your home. You can then estimate your wood needs or costs knowing that a cord of dry wood weighs about 4,000 pounds.

D—BTU/Hour Output. Perhaps the most important information for the consumer is listed in the second column. This figure shows the useful heat output into the room in Btu. If you buy a unit too big for the heat output required for your home, you will be operating it in an inefficient manner.

E—Percent Efficiency Range. The percent efficiency range is determined by dividing the energy flow into the room by the energy in the wood. These are the figures that will best enable you to compare units.

F—CFM Room Air Required. Column four lists the volume of room air consumed by the appliance in cubic feet per minute (cfm). Obviously, the more air that is consumed by the unit when the temperature is very cold, the less efficient the unit will operate.

		THE MAJESTIC COMPANY MODEL WM 36-1	(A)	
	WOOD LBS/HR	BTU/HR OUTPUT	% EFFICIENCY RANGE	CFM ROOM AIR REQUIRED
(B) 1. 2. 3.	(C) 13.7 16.3 18.7	(D) 43476 55780 53657	(E) 38-40 41-43 34-36	(F) 41 42 45

TEST UNIT EQUIPPED WITH ACCESSORY _____

These data were obtained under test conditions in accordance with the Fireplace Institute Standard I-79 for wood-fired, open combustion chamber heating appliances, and may not be reproducible in home operation

Fi
FIREPLACE INSTITUTE

Installing Wood Burning Systems   6

Wood is becoming a favorite alternative fuel. Attracted by the prospect of saving energy and money, thousands of homeowners all across the nation are installing wood burning appliances. Unfortunately, many of these systems are being improperly installed or carelessly operated, so wood appliance fires are becoming more and more common. Wood burning systems are not inherently dangerous; most of these fires could have been easily prevented by common safety measures.

Safety is not an exciting subject, but it deserves as much attention as if your life depended on it, for it well may. When installing a stove, you must use more than your common sense as your guide. Trial-and-error is no way to learn about safety. You must be right the first time.

Some homeowners have been so mesmerized by their new wood burning unit that they have neglected an extremely important factor—insurance. Unless you notify your insurance company of your new heating acquisition, you run the real risk of invalidating your whole policy. Insurance companies are aware that there has been a sharp rise in home fires related to wood or coal heating units. As a result, fire insurance companies have become very cautious in issuing new policies, or in renewing old policies to people who install wood burning systems. However, according to some insurance company officials, there is little problem if the unit bears the Underwriters' Laboratory seal and if the unit is properly or professionally installed. In any case, inform your insurance company immediately, even before installation.

To protect the individual and the community against home fires, cities, towns, and states have established a series of codes related to the installation and operation of heating systems, including wood. Check with your local fire department, fire marshal, town or city offices to obtain information regarding the installation of wood burning stoves or fireplaces. Observe any and all legal codes; failure to do so could lead to legal problems, court costs, and possibly a fine, to say nothing of running the risk of a home fire.

Your unit will come with specific instructions regarding its installation. You should follow these instructions to the letter or you run the risk of voiding your warranty. If there is a conflict between local codes and the manufacturer's specific instructions, follow the safest procedure. Thus, if the manufacturer suggests a 28-inch clearance between the unit and the wall, while local codes specify a 36-inch clearance, follow the latter code. In fact, most manufacturers recommend that when you are in doubt about anything, consult the local fire marshal for the latest and best advice.

Although most units come with clearly prepared instructions to allow a reasonably skilled do-it-yourselfer to install his/her own unit, there is no doubt that a professional job has its advantages. If nothing else, it will mean that you will not have any problems with your insurance company and that you may not subject yourself to legal difficulties with the local authorities. In addition, if you are the least bit uncertain about your ability to install a unit properly, you should have it professionally installed. But, whether you are going to do the installation, or have it professionally done, it is important that you know how it is done properly.

STOVE INSTALLATION

The most important consideration in installation is to ensure that the wood burning stove is properly installed for safe and efficient operation. It makes little sense to purchase a wood burning stove and through improper installation, lose or seriously damage a home through fire.

The first item to consider is location. If an existing chimney is to be used, its location and the length of the chimney flue may be determining factors. If a chimney is not available, then consideration should be given to installing an internal, factory-built chimney or exterior masonry chimney.

The second consideration (after selecting location) is to ensure adequate clearances between the stove and any combustible wall or home furnishings. A combustible wall is an unprotected wall containing wooden structural members—even though the wall may be surfaced with sheetrock, plaster, paneling, or other decorative facing. It is advisable to consult local fire and building codes for proper clearances and to determine if an installation permit is required.

The National Fire Protection Association (NFPA) has developed minimum clearance recommendations for combustible materials which serve as basic guidelines for stove installation. These guidelines are the basis for most local building codes.

These guidelines were developed prior to the current wood stove revival, and several stoves today have larger radiating surfaces than were common when the guidelines were promoted. Therefore, it is important to consult the owners manual which should be supplied with each stove and to check the stove clearances recommended by the manufacturer. Many stove manufacturers should have their stoves tested by a competent testing laboratory and the findings listed with one or several recognized building code groups. This way, if there is any variation between the NFPA recommended minimum clearances and clearances for the respective stove, the homeowner is in a position to adjust to the safe operating clearances. However, if the stove that you have purchased is not listed by a national testing laboratory (such as U.L.), follow NFPA recommendations given in this book.

Wall and Ceiling Protection

The sides of an unlisted radiant-type stove should be at least 36 inches from any combustible materials (Fig. 6-1A). A foot or two of clearance will not do. Even though very high temperatures are needed to ignite most combustible materials, over time high temperatures can change the composition of the material by slowly darkening it so that it accepts more and more radiated heat.

Installing Wood Burning Systems 121

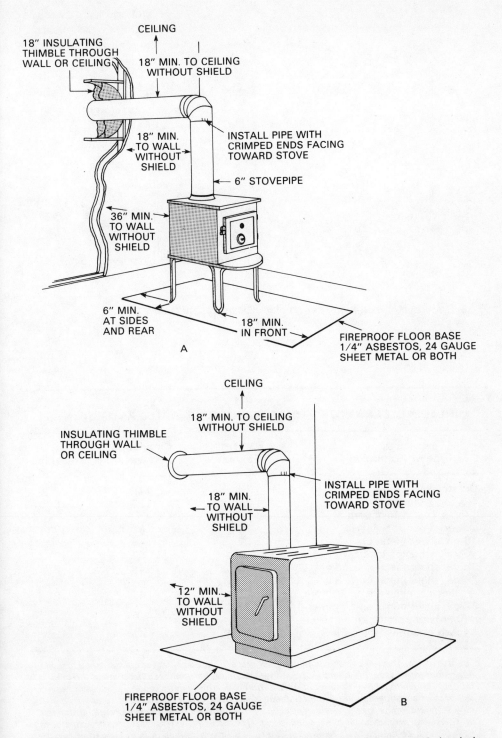

Fig. 6-1: Minimum clearances from unprotected surfaces for radiating and circulating woodburning stoves as recommended by the National Fire Protection Association (NFPA).

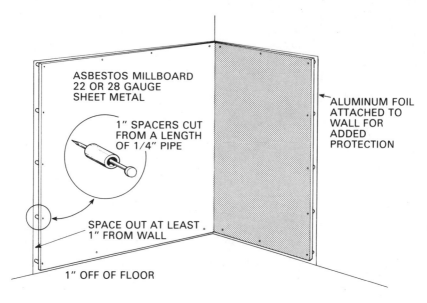

Fig. 6-2: Method of installing sheet metal or asbestos millboard.

Table 6-1:
MINIMUM CLEARANCES FROM COMBUSTIBLE WALLS AND CEILINGS

Type of Protection	Radiant Stoves	Circulating Stoves	Stovepipe
Unprotected	36"	12"	18"
1/4" asbestos millboard*	36"	12"	18"
1/4" asbestos millboard spaced out 1"*	18"	6"	12"
28-ga. sheet metal on 1/4" asbestos millboard	18"	6"	12"
28-ga. sheet metal spaced out 1"	12"	4"	9"
28-ga. sheet metal on 1/8" asbestos millboard spaced out	12"	4"	9"
22-ga. sheet metal on 1" mineral fiber batts reinforced with wire mesh or equivalent	12"	4"	12"

*The use of asbestos millboard in wall and floor protection is a controversial issue because of the health hazard of asbestos fibers in the manufacture, preparation, and handling of the millboard for use. The National Fire Protection Association is currently initiating the process of removing asbestos as a standard protection for reduced clearances. Since the process is a lengthy one, this new standard will probably not be in effect until the early 1980's. We strongly encourage use of an alternative protection whenever one is available. However, if you must use the asbestos millboard, use it cautiously. It is recommended to paint the asbestos to help keep the fibers from coming loose. If the board must be cut, do not inhale the dust; do the work outdoors, using a breathing mask.

Finally, it may begin to smolder and burn at temperatures as low as 200 to 250 degrees F. These temperatures are easily reached by an unprotected wall exposed to a stove without adequate clearance, so strict adherence to the clearance standards is advised.

As mentioned in Chapter 5, a double-wall circulating stove with an outer shell does not get as hot as the radiant-type and can be placed as near as 12 inches to an unprotected wall (Fig. 6-1B). As shown in Table 6-1, the recommended clearances can be reduced considerably if combustible walls and ceilings are protected with asbestos millboard or 28 gauge sheet metal spaced out 1 inch from the combustible wall (Fig. 6-2). The spacers should be constructed from a noncombustible material. Provide a 1-inch air gap at the bottom of the asbestos millboard or metal panel. Air circulating behind the panel will cool the panel and the wall.

Asbestos millboard is a different material from asbestos cement board. Asbestos cement board (transite) is designed as a flame barrier; it provides little in terms of heat resistance—it will conduct heat to any combustible surface to which it is attached. Brick and stone also provide little or no protection for a combustible wall because they are good conductors of heat. To be effective, bricks must be spaced out 1 inch from the wall with air gaps at the top and bottom (Fig. 6-3). This can be accomplished by using half bricks in the top and bottom rows.

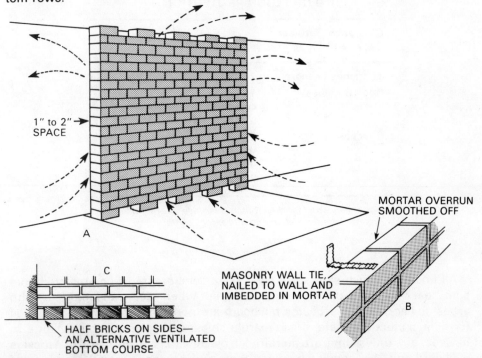

Fig. 6-3: (A) The protective barrier spaced away from the wall, allowing air circulation. It is not necessary to extend the barrier to the ceiling unless the size of the stove dictates it. (B) How the spaced barrier is fastened to the wall by means of a masonry wall tie. (C) An alternative base arrangement of the bricks. The floor protection should extend all the way to the main wall.

A dry wall (gypsum wallboard) over studs is considered a combustible wall. Heat is transmitted directly through the dry wall to the studs. Remember that clearances mentioned here are for all combustible materials. This eliminates some dangerous but common practices such as stacking wood or paper next to a stove or moving a sofa "temporarily" closer. It may be difficult to install a stove where it is convenient, aesthetically pleasing, and safe, but do not compromise. Your priority is safety.

Floor Protection

Safe floor clearances are smaller in size than those for walls because the heat radiated from the bottom of a stove is generally less than from either the sides or the top. During a fire, ashes fall to the bottom of a stove; this has an insulating effect, resisting the flow of heat downward. This process may be started in a new stove by placing a layer of sand in the bottom. Clearances for proper floor protection are classified according to stove leg length and are listed in Table 6-2.

**Table 6-2:
NFPA RECOMMENDED CLEARANCES
FOR FLOOR PROTECTION**

Clearance Between Stove and Floor	Protection Needed
18 inches or more	24-gauge or thicker sheet metal
6 to 18 inches	1/4-inch asbestos millboard covered with 24-gauge sheet metal
6 inches or less*	4-inch thick hollow masonry units with holes interconnecting and open, covered with 24-gauge sheet metal

*For stoves with less than 6-inch clearance, see manufacturer's instructions for building up a fireproof platform.

All floors on which stoves are set, except concrete, must be protected from both heat of the fire and hot coals falling out when fuel is added. Metal with asbestos backing and asbestos millboard are noncombustible materials used for floor protection. Slate, brick, marble chips, and colored pebbles can also be used; but, unless they are mortared in place with no gaps, metal or asbestos millboard must be installed between them and a wood floor. A 2-inch layer of sand or ashes, or bricks laid in the bottom of the stove, helps prevent overheating of combustible flooring. When using any of these noncombustible materials, they must be used in conjunction with asbestos millboard sheet or sheet-metal plate (Fig. 6-4).

Installing Wood Burning Systems 125

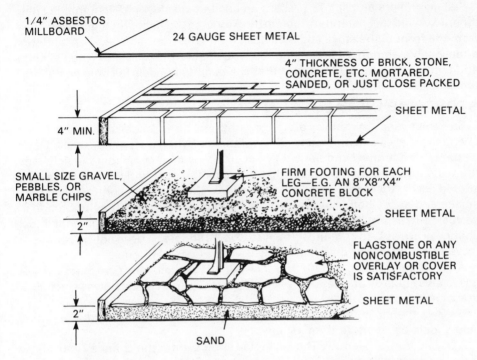

Fig. 6-4: Four floor protection plans for wood stoves with legs 6 inches long.

In recent years, several manufacturers have come out with wall panels and floor pads of real or simulated brick. They are designed to mount flush on the wall or to be placed on the floor with the proper clearances between combustible material and the stove (Fig. 6-5). Before using any of these new wall panels or floor pads, check their ratings or their approval with your local fire code.

Fig. 6-5: Typical simulated brick wall panel and floor pad. The simulated hollow-core brick panel and pad have a very low heat transfer rate. Fiberglass insulation in the core further minimizes heat transmittal.

Falling embers and sparks present an additional safety problem that is often ignored. Avoid this potential problem by extending the floor protection 18 inches from the front of the stove and 6 inches around the sides and back. This affords a reasonable amount of protection, but you should still take care when loading and tending the stove. Make sure that ashes and hot coals fall only on the protected area.

STOVEPIPE INSTALLATION

Stovepipe and chimneys, although they are often mistakenly used interchangeably, are two completely different things. When a stovepipe is used for a chimney, dangerous conditions may occur. Creosote may build up rapidly, and the wind and rain may soon corrode the pipe. Consequently, if a chimney fire should occur, there will be little to contain it. Chimneys, in contrast, keep the smoke and gases hotter to prevent the buildup of creosote. They are made of corrosion-resistant materials, so they do not need frequent replacement. Most chimneys are able to contain a fire. A stovepipe should be used only to connect a stove to a chimney and should never be used instead of one.

A stovepipe must be installed so that there is a good "draft" to carry the hot gases away quickly and safely. This can be a tricky proposition unless you follow these guidelines carefully.

1. Keep the stovepipe as short and straight as possible. The total length of pipe should be no more than 10 feet.

2. Avoid horizontal runs. The stovepipe should enter the chimney well above the stove outlet. Horizontal portions of the stovepipe should rise at least 1/4 inch per foot.

3. The horizontal portion of the stovepipe should be no more than 75 percent of the vertical portion.

4. Keep turns and bends to a minimum. There should be no more than two 90-degree elbows.

5. Never run a stovepipe through a window; this is a dangerous practice. Also, do not run through concealed places like closets or attics.

6. The minimum safe clearances are specified in Table 6-1, but the owner's manual should be consulted for specific stovepipe clearances.

All stoves have some sort of damper. Air inlet dampers are used to control the air flow to a stove. In the event of a chimney fire, it should be possible to shut off the air flow to the chimney completely. Airtight stoves have this capability, but non-airtight stoves can benefit from having a damper in the stovepipe. The common damper can significantly reduce the air flow to the chimney and thus reduce the intensity of a chimney fire.

Stovepipe is sold in inch diameters at various foot lengths. Most building codes require stovepipe to be 24 gauge or thicker; lower gauge numbers indicate thicker metal. The diameter of the stovepipe used should be the same diameter as the firebox outlet. Most wood stoves use either a 6- or 8-inch smokepipe. Using stovepipe that is smaller in diameter than the firebox outlet will reduce combustion efficiency and possibly cause improper draft. If your stove is a foreign make, it will usually require metric size pipe. Should you be unable to purchase metric stovepipe, adapters are available to match inch sized sections.

Sections of stovepipe should be joined so that the crimped or inner end is closer to the stove. In that way, any creosote that forms will drop down the pipe into the stove, rather than spilling out at a stovepipe joint. The stovepipe sections should be secured together with three equally spaced sheet metal screws. The pipe should be supported by wire or metal hangers if the horizontal length of stovepipe exceeds 6 feet. Since stovepipe cannot be expected to last forever, inspect it regularly and plan on replacing it every two or three years. (The sulfuric acid which forms in the smoke condensate will rust out the stovepipe, resulting in holes which may be dangerous if undetected.)

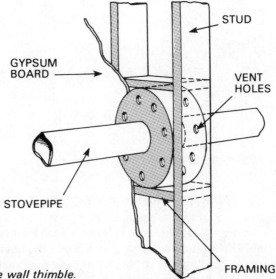

Fig. 6-6: Ventilated-type wall thimble.

Passing a stovepipe through an interior wall should be avoided if possible. If it cannot be avoided, special precautions must be taken. An insulated wall thimble may be installed (be sure it is nationally listed for use with wood stoves) according to manufacturer's instructions, or a ventilated thimble (Fig. 6-6) that must be specifically made for wood stoves can be used. This thimble should be made of sheet metal or asbestos millboard and must be at least three times larger in outside diameter than the stovepipe. For a 6-inch stovepipe, use a thimble that is 18 inches in diameter.

Stovepipe, when it is fit into an outside chimney, should extend into the chimney only as far as the inner edge of the flue liner. The pipe should be securely cemented to the chimney using high-temperature furnace cement.

There are four ways to safely pass a stovepipe through a combustible wall in order to hookup with a chimney flue:

1. Use a ventilated thimble (Fig. 6-6) that is at least three times larger in diameter than the stovepipe.

2. Use a U.L. ALL FUEL or CLASS A thimble (Fig. 6-7A) extending through the wall, with a wall hole 2 to 4 inches larger than the thimble diameter. This permits the placement of an insulating material such as fiberglass or rock wool between the thimble and the wooden framing of the wall. Cover the gap between the wall and the stovepipe with a stovepipe flange.

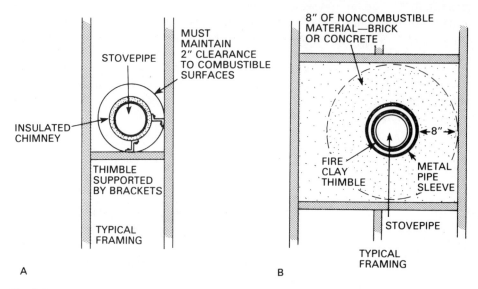

Fig. 6-7: Two other types of wall thimbles and how they are installed.

3. Use a burned fire-clay thimble (Fig. 6-7B) and surround it with at least 8 inches of fireproofing material such as fiberglass insulation or brick. Cover the opening with noncombustible materials such as asbestos millboard or metal. A small gap should be left between the thimble and the covering material to allow either the house or chimney to settle slightly and not crack the thimble. The gap can be covered with a stovepipe flange.

4. Use no thimble whatsoever, but securely fasten the stovepipe to the chimney with a high temperature cement. Combustible material within 18 inches of the pipe must be removed. For a 6-inch diameter pipe, this requires a 6 inch + (2 x 18 inch) = 42-inch diameter hole in a combustible wall. The hole may be closed in or covered with noncombustible materials such as masonry, asbestos millboard, or sheet metal.

When the wall is cut between supporting studs for the thimble, inspect the opening to make sure there are no electrical wires or conduit in the space between adjoining wall studs. Heat from the stovepipe may be sufficient to melt the insulation on wire in this space, causing an electrical fire.

Stovepipe Damper. Since the flow of air into an airtight stove can be controlled by the stove vent, a stovepipe damper is not needed. The exception is when an airtight stove is overly affected by too much draft during windy periods; a stovepipe damper will reduce the pull of heavy drafts. With a non-airtight stove, a stovepipe damper is a must.

The most popular manual stovepipe damper is the so-called "twist" type, which consists of a cast-iron damper plate or disk and a metal spring-loaded handle and rod. The handle rod detaches under pressure from the disk. The damper should be located at the lowest operable point on the first section of stovepipe—usually about 6 inches from the stovepipe collar on the stove (Fig. 6-8A).

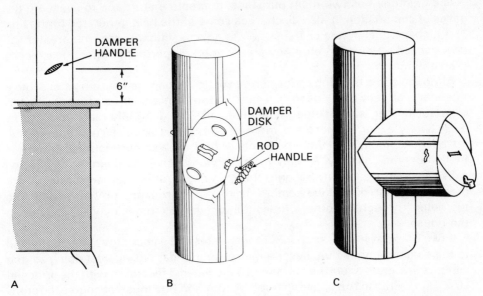

Fig. 6-8: (A) Location of damper; (B) parts of a damper; and (C) a barometric draft control.

To install a twist damper, first drill a hole in the side of the pipe the size of the metal handle rod. Holding the damper disk inside the stovepipe, insert the rod through both the hole in the side of the pipe and disk until it strikes the other side of the pipe. Once pinned check the positioning of the damper disk by turning the handle to see that it moves freely (Fig. 6-8B). Once the damper position is set, push on the rod until it dimples the other side of the pipe. Then, drill the second hole on that dimple. With the damper disk in position, push the rod through the second hole and lock the plate onto the handle pin.

To operate a twist damper, push the handle in, turn the handle in the desired direction, and then pull the handle out to hold the disk in position. The handle of damper will tell you the position of the damper in the pipe. In a vertical stovepipe arrangement, a horizontal positioned handle would indicate that the damper is closed; a vertical or angled handle position signals that the damper is open. For stoves with a horizontal exit pipe, a handle in a vertical position means that the damper is closed, while a horizontal or angled handle position shows that the damper is open.

Barometric draft controls (Fig. 6-8C) are available and can be installed in place of the manual stovepipe damper just described. This control opens automatically if the draft becomes too high; the manual type must be closed to achieve the same results. Both decrease efficiency since (1) they are not airtight and (2) they do syphon off heated room air when operating.

CHIMNEYS

Chimneys present an uncertain danger. Most of them are concealed or inaccessible, yet they play a critical role in the overall safety of your wood burning system. Be sure that your chimney will safely do its job.

The chimney has two main purposes: to create a draft and to evacuate the gases of combustion. It also discharges some of the heat generated by the fire. The higher the chimney or the larger its cross-sectional area, the greater the flow capacity. However, chimney area is more important in affecting capacity than chimney height.

Although some building codes permit you to vent a wood burning stove into a flue that is already being used by a gas or oil appliance such as a furnace, boiler, or water heater, shared-flue installations are not generally recommended by most safety experts. There are three good reasons against such shared-flue installations. First, each time the furnace shuts off, a small amount of unburned fuel enters the chimney. A spark from the wood stove could ignite the gas and cause a small explosion. Second, the chimney is often not large enough for proper operation of the two heaters. Third, gases from one unit may come into the house through the other unit so that dangerous fumes may accumulate in the house.

If two or more stoves, such as a room heater and a cook stove, are connected to the same chimney flue, despite the recommendations against doing so, the connections must enter the chimney 18 inches to 3 feet apart with the principal stove connected to the upper opening. A common flue must, of course, be sufficient size to provide an adequate draft for all the stoves connected to it.

If you are lucky enough to have an existing, unused, and appropriately sized and located chimney that is in good condition, it can usually be used. If you do not have one available, you will have to install a new chimney.

New Chimneys

There are two types of new chimneys in common use: prefabricated metal and masonry.

Prefabricated Chimneys. Prefabricated chimneys are easier to erect than masonry ones. Tests at the National Bureau of Standards have shown that metal and masonry chimneys differ little with respect to draft when used under similar conditions. A key point is that metal prefabricated chimneys must be U.L. listed as ALL FUEL or CLASS A chimneys. Under no circumstances should a prefabricated chimney carrying a Type B Gas "vent" be used, since it is not designed to handle the temperatures generated with wood burning.

Figure 6-9 shows the three common types of prefabricated metal chimneys. The most popular type is the insulated one because it effectively keeps heat inside the flue. This action tends to reduce creosote accumulation in the flue, but it also reduces the heating efficiency of interior-chimney installations. The air-insulated type is best for interior-chimney locations. The triple-walled, air-cooling "thermal syphon principle" prefabricated chimneys (Fig. 6-9C) should only be used where creosote buildup will not be a problem. This design of metal chimney seems to have a greater creosote buildup than the others.

Details on both exterior and interior installations of prefabricated chimneys are given later in this chapter.

Masonry Chimneys. Masonry chimneys are more costly to construct, but they have benefits that may outweigh their high cost. For example, a chimney built inside the house has the capacity to store excess heat and temper the inte-

Installing Wood Burning Systems 131

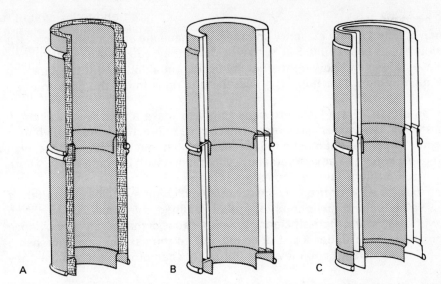

Fig. 6-9: Three types of prefabricated metal chimneys: (A) Insulated; (B) air-insulated; and (C) air ventilated.

rior environment. That is one of the reasons why in Colonial New England it was common practice to build houses with large center chimneys. These inside masonry chimneys increase the net efficiency of the heating system by up to 15 percent due to heat "leaking" from the chimney into the house. In addition to being more attractive, masonry chimneys also tend to be more durable because they are more resistant to corrosion. But, remember that a masonry chimney takes time to warm up, thus causing a small amount of creosote to be deposited inside the flue every time a fire is started.

A new masonry chimney can be built of brick, stone, or special concrete chimney blocks (Fig. 6-10). When building up the chimney, be sure that there is an air space between the masonry material and the tile flue liner. Never fill this space with mortar. The flue liner should extend at least 8 inches below the entry of the stovepipe. The flue liner should be cemented together with a medium-duty refractory mortar. Check local building codes for minimum clearances (usually 2 to 3 inches) between combustible materials and other construction details before starting a new masonry chimney project.

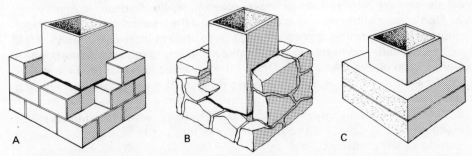

Fig. 6-10: Three common types of masonry chimneys; (A) Brick; (B) stone; and (C) prefabricated concrete chimney blocks.

Old Chimneys

When building a new masonry chimney, you have the opportunity to check it for safety as it is being built. But, with an old chimney, you have no such luxury. You have no way of knowing the quality of the initial job, and it is difficult to check its durability over time. The best thing to do is to get the opinion of an expert.

Ask an official of the local fire department to make a chimney inspection. Generally, the department welcomes the chance to practice some "preventive medicine." A competent mason may be able to point out defects, and the local building inspector can advise you about compliance with building codes. There are also some things you can check (see Chapter 8).

First, look for a tile or fire-clay liner. Not all old chimneys have one, but their proven safety value makes them essential. Unlined chimneys should not be used with airtight and high-efficiency stoves because of the increased incidence of creosote buildup under low fire conditions. In the case of severe creosote buildup and associated chimney fires, flue liners can protect against chimney burnout.

To check the condition of the liner, look up or down into the chimney flue with a flashlight; the flue should look smooth and uniform. Scrape its surface and you will find that it is much harder than brick. The flue tile liners are brittle, and they sometimes crack. You must be sure that the liner is in good condition, so that hot gases and creosote cannot escape the flue prematurely and start a fire. Visually check for cracks, then perform a smoke test. Build a smokey fire in the fireplace or stove, and then cover the top of the flue. This will force the smoke through any cracks or holes. Leaks must be repaired or the chimney should not be used. Check for loose mortar by scraping between the bricks. If the mortar crumbles easily, you will need to have the bricks repointed, the old mortar scraped out, the brick removed, new mortar installed, then the brick replaced.

For chimneys with no liner, it is possible to have metal liners fabricated and inserted into the chimney (Fig. 6-11). It is advisable to have these liners fabricated from 20 gauge (.0375 inch) stainless steel in grades 304 or 316. This type of liner should give long life and excellent service, in contrast to light gauge steel liners that would deteriorate in a short period of time.

There are other "don'ts" to be aware of when checking an old chimney. For instance:

1. **Don't** use a masonry chimney that is supported by brackets or shelves. The brackets may weaken and the chimney may fall while in use.

2. **Don't** use a chimney that is used to support the framing of the house. Heat conducted to the framing may eventually start a fire. For interior chimneys, a full 2 inches of clearance between the chimney and any combustible materials is required. Zero clearance is acceptable if the chimney is 8 inches thick at that point. An exterior chimney may have one side in contact with exterior sheathing in some jurisdictions.

3. **Don't** use snap-in covers to close unused smoke pipe inlets. A chimney fire can easily pop them out. Fill the hole with masonry or tile built up to the same thickness as the chimney.

4. **Don't** use a chimney whose flue area is more than twice as large as the proper size stovepipe for your stove. This will help prevent backdrafts.

Installing Wood Burning Systems   133

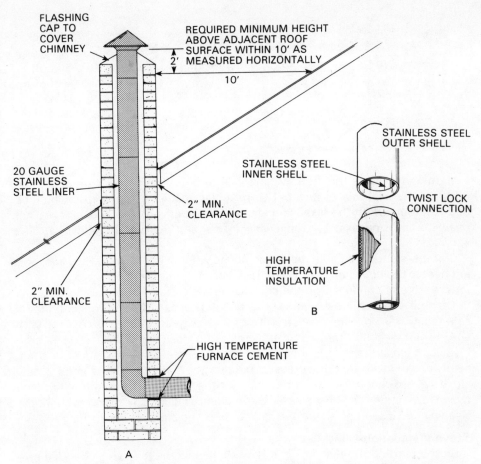

Fig. 6-11: (A) Things to look for when inspecting an old masonry chimney. (B) How liner is installed in an existing chimney.

National Fire Protection Association Standards state that a stove chimney connector is not permitted to be connected to a flue serving a fireplace—a fireplace must have its own individual flue. Franklin stoves have an open front and should be treated as fireplaces in this respect.

The final area for inspection is chimney height. A chimney should extend at least 3 feet above flat roofs. On pitched roofs, chimneys should be 2 feet higher than any point within 10 feet, to prevent downdrafts and fires from sparks.

The flue lining of a masonry chimney is extended 4 inches above the top course of brick or stone and the top of the chimney capped with cement mortar. The mortar is 2 inches thick at the outside edges of the chimney and sloped up to the flue lining to direct air currents upward at the top of the flue and to drain water from the top of the chimney. It is a relatively simple task for a mason to extend a chimney that does not meet these minimum standards.

A chimney cap is sometimes used to help prevent downdrafts where the chimney's top is subject to wind turbulence caused by roof shape, trees, terrain, or other buildings and to keep out rain and snow. Any cap adds resistance to the

SCREEN CAP   DISK CAP   SPINNING CAP   SWEEPSTACK

Fig. 6-12: Examples of chimney caps.

system and reduces the draft. Mechanical turbines, revolving ventilators, and other mechanical devices (Fig. 6-12) are subject to failure from creosote buildup and weather. Often the disadvantages outweigh advantages and caps are not used. If a cap is necessary, a removable flat disk cap is simple and slows gas flow very little.

Chimneys should be built inside the house whenever possible. They add radiant heat to the house and tend to remain warmer, thus creating less opportunity for creosote to condense on the chimney walls.

Keeping a safe chimney should be a constant concern. Even if your chimney is safe today, this does not mean it will be safe next year. Chimneys get tired with age, mortar weakens, they settle and crack, and loose bricks, bird nests, or beehives can block them. Frequent inspections are necessary to insure adequate safety and to make sure that creosote buildup is within reasonable limits. A new stove system should be checked weekly for the first month. Length of time between inspections can be increased as experience dictates. A thorough chimney-cleaning and inspection should be made prior to the start of each heating season.

Stove Installation Checklist

Before starting the first fire in a new stove, use the following checklist prepared by the Northeast Regional Agricultural Engineering Service to be certain that it is safely installed:

- ☐ 1. The stove does not have broken parts or large cracks that make it unsafe to operate.
- ☐ 2. A layer of sand or brick has been placed in the bottom of the firebox if suggested by the manufacturer.
- ☐ 3. The stove is located on a noncombustible floor or an approved floor protection material is placed under the stove.
- ☐ 4. The stove is spaced at least 36 inches away from combustible material. If not, fire-resistant materials are used to protect woodwork and other combustible materials.
- ☐ 5. Floor protection extends out 6 to 12 inches from the sides and back of the stove and 18 inches from the front where the wood is loaded.
- ☐ 6. Stovepipe of 22 or 24 gauge metal is used.
- ☐ 7. The stovepipe diameter is not reduced between the stove and the chimney flue.
- ☐ 8. A damper is installed in the stovepipe near the stove unless one is built into the stove.
- ☐ 9. The total length of stovepipe is less than 10 feet.

- ☐ 10. There is at least 18 inches between the top of the stovepipe and the ceiling or other combustible material.
- ☐ 11. The stovepipe slopes upward toward the chimney and enters the chimney higher than the outlet of the firebox.
- ☐ 12. The stovepipe enters the chimney horizontally through a fire-clay thimble that is higher than the outlet of the stove firebox.
- ☐ 13. The stovepipe does not extend into the chimney flue lining.
- ☐ 14. The inside thimble diameter is the same size as the stovepipe for a snug fit.
- ☐ 15. A double walled ventilated metal thimble is used where the stovepipe goes through the interior wall.
- ☐ 16. The stovepipe does not pass through a floor, closet, concealed space or enter the chimney in the attic.
- ☐ 17. A U.L. approved ALL FUEL or CLASS A metal chimney is used where a masonry chimney is not available or practical.
- ☐ 18. The chimney is in good repair.
- ☐ 19. The chimney flue is not blocked.
- ☐ 20. The chimney flue lining and the stovepipe are clean.
- ☐ 21. A metal container with tight fitting lid is available for ash disposal.
- ☐ 22. The building official or fire inspector has approved the installation.
- ☐ 23. The company insuring the building has been notified of the installation.

INSTALLING A NEW FIREPLACE

At one time, building your own fireplace was a very ambitious project, but today, thanks to prefabricated fireplace units and chimney kits (Fig. 6-13), it is a relatively simple one. Before starting any fireplace project, consult the local building code, especially the requirements for installation of factory-built fireplaces and chimneys.

As mentioned in Chapter 5, there are two basic types of prefabricated fireplaces: the zero clearance style and the free-standing style.

Fig. 6-13: Typical prefabricated chimney kit.

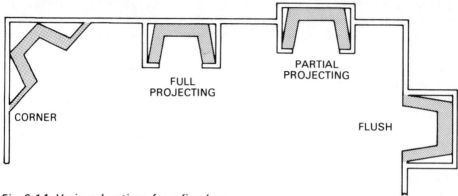

Fig. 6-14: Various locations for a fireplace.

Zero Clearance Fireplaces

There are several ways that the zero clearance fireplace can be installed, as shown in Fig. 6-14. Consider the traffic pattern in your room and check the con-

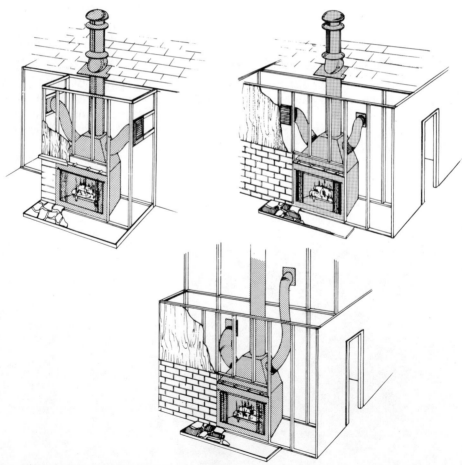

Fig. 6-15: Methods of locating hot air ducts.

struction of your home above and below the fireplace before making a choice. A corner location may be best where space is limited.

The fireplace may be installed directly on the floor, or an elevated wood or masonry platform may be used. When placed on a combustible floor, a fireproof hearth extension is sometimes required. (This may be purchased with the fireplace unit.) The extension may be on the same level as the fireplace, or it can be recessed into the floor. Both the fireplace and the extension should be on a flat, level surface. A projecting fireplace will accent large natural finishing materials such as fieldstone.

When planning the location of a zero clearance fireplace unit, consider the location of the grille system. As shown in Fig. 6-15, the grilles can be located in the sides of the fireplace, the front of the fireplace and an adjoining room, and an adjoining room and an upstairs room.

The total height of the fireplace and chimney, in most cases, should not be over 45 feet. The installation will be easier and cost less if joists or rafters are not cut as the chimney is assembled. The simple measurements are indicated in Fig. 6-16. A 15-degree elbow is available with most units, and it can be used to offset the chimney to miss a rafter or joist.

The fireplace may be positioned and then the framing built around it (Fig. 6-17), or the framing may be constructed and the fireplace then pushed into the opening. The fireplace may touch combustible materials at the bottom, sides, and back. Chimney sections must have at least a 1-inch clearance to combustible material. Firestop spacers must be installed at every ceiling level. They will provide the necessary clearance. Install the fireplace no closer than 36 inches to any unprotected combustible wall, perpendicular to the fireplace openings. This does not apply to corner installations or walls with insulated shields. A floor that supports a fireplace does not have to be reinforced unless the material used for facing is very heavy, i.e., large areas of brick, fieldstone, etc.

As you install your fireplace, you may want to include space for a face finishing material. You can use plasterboard, plywood, or wood paneling, ceramic tile, or any similar material. The material or facing may be installed even with the wall, or it may project in front of the wall. If you want the facing and wall to be even, you must recess the fireplace unit back the thickness of the facing.

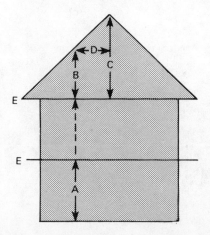

Fig. 6-16: Measurements needed to figure a fireplace installation: (A) The height of the room in which the fireplace will be installed, plus the height of any rooms overhead through which the chimney must pass. Note, if the fireplace is to rest on a raised platform for raised hearth effect, deduct the height of the platform from your total measurement. (B) The height from the attic floor to the point through which the chimney will exit through the roof. (C) The height of the peak of the roof from the attic floor. (D) The horizontal distance from line B to C. (E) The number and thickness of ceilings through which the chimney will pass. With this information, you can select all of the necessary components.

The hearth or the area in front of the fireplace should be of fireproof material. If the floor is wood, you can cover it with bricks, ceramic tiles, sheet asbestos, or metal. Typically, a hearth should be 12 to 20 inches deep and extend about 8 to 10 inches on either side of the opening. If you prefer the fireplace and hearth raised above the floor, you can get a raised base or you can build your own platform out of wood. If the fireplace is on a platform, you may bring the facing around the bottom of the opening. Do not block off the air opening at the bottom of the fireplace.

Figure 6-18 shows the method of installing a metal fireplace with a prefabricated chimney.

Free-standing Fireplaces

Free-standing prefabricated fireplaces are the easiest to install. Most building codes specify that the back and sides of the free-standing fireplace be kept a certain distance from combustible materials, unless it has a high-temperature insulation barrier built into it, as is the case in models designed for wall installa-

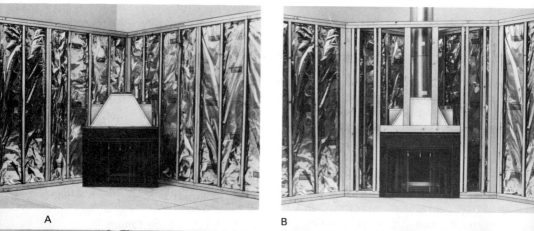

Fig. 6-17: Framing details for a built-in zero clearance corner fireplace.

Installing Wood Burning Systems 139

tion. Most models must be set at least 18 inches away from the wall. The unit should stand on a base made of a noncombustible material such as brick, metal, or tile in a manner similar to that shown in Fig. 6-4.

Fig. 6-18: Method of installing a fireplace with a prefabricated chimney.

Once the fireplace has been assembled according to the manufacturer's instructions, locate it where you plan to make the complete installation. To obtain the most efficient fireplace, built-in or free-standing, it should be located as close to the chimney as is practical (Fig. 6-19). Keep the smokepipe connection to the chimney flue short and direct, using as few elbows as possible. When connecting the stovepipe between the fireplace and chimney, follow the same basic precautions given earlier for stoves.

The installation of a prefabricated chimney for either a free-standing unit or built-in fireplace generally involves the following steps (Fig. 6-20):

1. Cut an opening through the ceiling for the support box. Try to locate the center of the opening so a joist will not be cut. Cut out the opening as directed by the manufacturer. Then, install headers between the joists for extra support.

2. Install the chimney support box. Place the metal support box up through the opening until the bottom is against the ceiling. Excess metal above the joists may be cut off. With the box or collar in position, nail it securely to the joists. If the chimney goes through an upstairs room in a corner or along a wall, it should be covered with a partition. This will help avoid personal injury as well as protect the chimney. Be sure to keep at least a 1-inch space between the chimney and the partition.

3. Cut an opening through the roof. Locate the center of the opening to avoid cutting a rafter or beam. If this cannot be avoided, even by using an elbow, headers must be placed between remaining rafters or beams to help support the roof members. Make the cut to the size recommended by the manufacturer. Remember that roof openings should be at least 1 inch larger, to keep the 1-inch minimum clearance from the chimney pipe to combustibles. When making a cut, save the shingles for covering the flashing.

4. Run the chimney smokepipe from the fireplace to the roof support box. After the roof support box is in place, the chimney smokepipe or wall pipe can be installed up to the roof. The pipe must be kept 1 inch from combustibles. The support box and the roof flashing act as spacers. Firestop spacers must be used any time you go through another floor.

5. Complete the chimney top installation. Once the chimney or wall pipe has been brought through the roof, the chimney top should be installed as directed by the maker. Keep in mind that the chimney outlet must be at least 3 feet above the roof cutout. It also must be at least 2 feet above the highest point of the roof within 10 feet in a horizontal plane. Flash the chimney unit as directed and then renail the shingles on the top and sides with roofing nails. Do not nail through the lower portion of the flashing above the shingles. Chimney hoods are made of aluminum painted a neutral gray or with a simulated-brick finish.

6. Connect the fireplace to the chimney. Align the flue collar of the fireplace with the chimney pipe. The last section of the chimney pipe is installed by using an adapter between it and the fireplace flue pipe.

When a chimney flue is installed through a wall or window to an outside chimney installation, a special tee is used. For best operation, the chimney smokepipe should not run horizontally for a distance exceeding 1/2 the vertical height of the chimney serving the fireplace. When the pipe goes through an outside wall, the horizontal flue should be pitched at least 1/4 inch for every foot of pipe used between the fireplace and the tee. This way you can be assured that the

Installing Wood Burning Systems 141

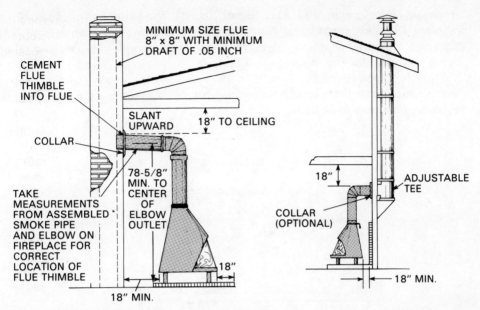

Fig. 6-19: (Left) Proper venting into an existing masonry chimney; (right) using a prefabricated metal chimney. Required clearances may vary with existing local building codes or manufacturer's instructions.

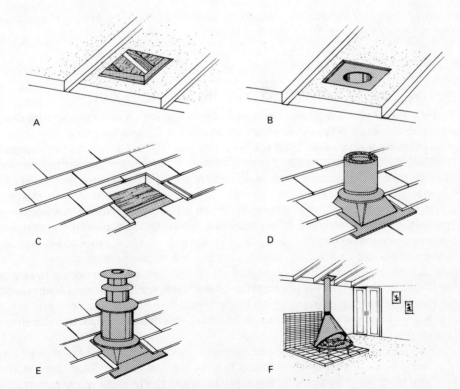

Fig. 6-20: Steps in installing a prefabricated chimney for a free-standing fireplace.

smoke will flow upward and not become trapped in the outlet and cause a smoke problem. Details of the outside installation of a chimney are shown in Fig. 6-21. Your heating supplier will help you determine the amount of piping and other items you need for the chimney installation.

Other fireplace items such as fireplace hoods are available in prefabricated form. If you follow the manufacturer's directions for assembling them, they are usually quite easy to install.

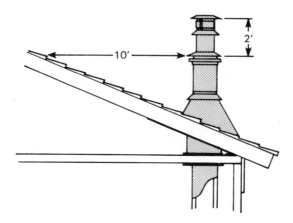

Fig. 6-21: The top of the chimney should be at least 3 feet above the roof and at least 2 feet higher than any point of the roof within 10 feet. It is always wise to extend the chimney above the peak.

INSTALLING A WOOD STOVE IN A FIREPLACE

While there are two types of stoves designed specifically for installation with a fireplace, almost any wood stove can be connected to a fireplace. The advantage of using a stove designed for fireplace installation is that it is sold with a fireplace cover or stoveboard. Whether installing a stove designed to fit inside the fireplace or one that is to be placed in front of the fireplace (see Chapter 5), the manufacturer's installation instructions should be followed to the letter.

Figure 6-22 shows three popular methods of installing a stove in a fireplace. The method shown in Fig. 6-22A requires a stoveboard to seal off the entire fireplace opening. A length of stovepipe is then run from the stove into an opening in the stoveboard or fireplace cover. This cover can be made from 1/4-inch asbestos millboard or a 1/8-inch piece of steel. It should be large enough to cover the fireplace plus allow for a 3-inch lap on each side and at the top. To determine the height of the stovepipe opening in the fireplace cover, measure the distance to the top of the stove's flue exhaust collar, then add to that 1/4 inch for every foot of stovepipe.

As shown in Fig. 6-23, there are several ways of holding the fireplace cover in place. The easiest method is to use two lag screws on each side of the cover. This entails drilling holes in the masonry between the bricks, inserting masonry anchors in the drilled holes, then drilling through the fireplace cover so that the

Installing Wood Burning Systems 143

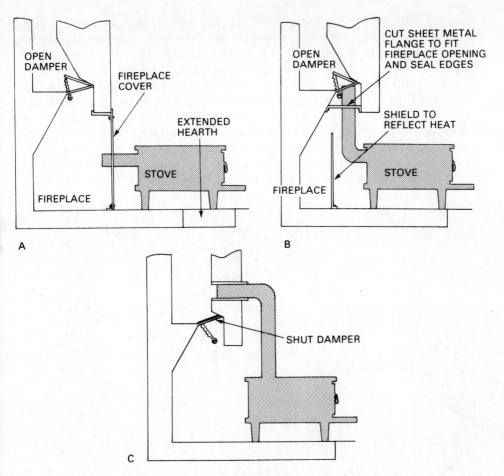

Fig. 6-22: Three methods of installing a stove in a fireplace.

lag screws will be embedded in the anchors. Of course, if you do not wish to deface the front of your fireplace, you can use either of the "L" bolt methods shown in Fig. 6-23B and C. But, before installing the fireplace cover, be sure to remove or wire open the damper and make certain that the ash pit is closed. Also, spread a layer of sand over the inner hearth to catch any creosote drippings. Also, before screwing the cover into position, run a bead of nonflammable caulking or stove cement around the edge of the cover to give a tight air seal on the irregular surfaces of the fireplace facing.

As shown in Fig. 6-22B, the stovepipe can be run directly into the fireplace and up to the flue. While this installation requires an elbow and one or more sections of stovepipe, it is usually easier to do than installing a fireplace cover. The insert piece for stovepipe is made from a piece of 18 gauge sheet metal and cut to fit the fireplace or damper opening. Use fiberglass insulation to seal the edges of the sheet metal flange and furnace cement to seal around the stovepipe as it goes through the flange. Be sure, however, that the stovepipe can be easily removed for cleaning. In some instances, creosote may run down toward the fireplace into the sealing material or past the damper and accumulate in the original fireplace opening. Check frequently for creosote accumulation. The material used to create the seal may need to be replaced periodically.

If the stove is too tall for the stovepipe to be inserted directly into the fireplace, or if you want to obtain heat from the stovepipe, the connection can be made above the fireplace (Fig. 6-22C). When making a fireplace installation, remember that if the stovepipe is within 18 inches of a wooden fireplace mantel or any other woodwork on the fireplace, a heat deflector shield of sheet metal or asbestos millboard should be constructed to protect the wood. Also, if the stove is set on the hearth, but the wood loading door opens over the floor, it is wise to place a piece of heavy sheet metal over the floor to act as a spark shield.

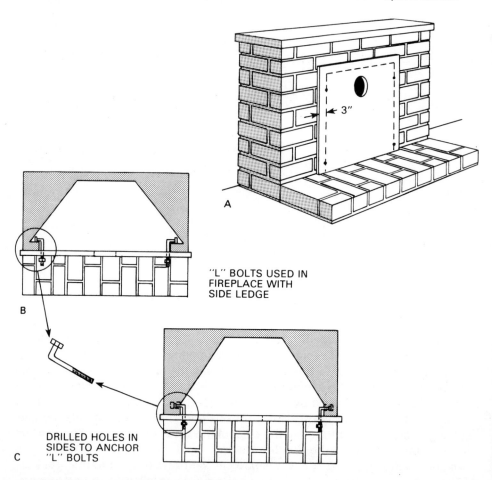

Fig. 6-23: How to install a fireplace cover.

Operating Wood Stoves and Fireplaces 7

After the wood stove or fireplace is installed, you are ready to receive its benefits—**heat**. Let us first look at how to get the most from your wood burning stove by operating it properly.

OPERATING A WOOD STOVE

A wood stove is not like a conventional oil or gas heating system. It cannot be set and forgotten; there is no electronic wizardry to do the thinking for you; it needs periodic attention. Still, most people consider this part of the lure of heating with wood, and they thoroughly enjoy the stoking, poking, and tending that their stove demands. So, operating a stove is not as simple as setting a thermostat, but it is a subtle blend of Science and Art. The Science is the thing we will teach you here, and the Art, well, that is something only time and practice can teach, but we are sure you will enjoy the process of discovery.

Three essential ingredients are required to begin the combustion process in a stove. Obviously, you need a match to start the fire and high temperatures to keep it going, but you also need fuel and air. Learning to operate a stove is simply learning to control these three variables: fuel, oxygen, and temperature.

The Fuel. The heat that is derived from the combustion of wood depends upon the concentration of: (1) woody material; (2) resin; (3) ash; and (4) moisture content. The first three features vary depending on the tree species and its growth rate, while the latter depends on the species, the season in which the tree was cut, and the seasoning procedures used. In general, the heaviest woods, when seasoned, have the greatest heating value. Lighter woods give about the same heat value per pound as heavier hardwoods, but, because they are less dense, they give less heat per cord or cubic foot.

The moisture in the wood is also an important factor affecting fuel quality. Burning wet wood wastes energy, because moisture in wood must be evaporated before it will burn, and evaporating this moisture takes energy. Naturally, the wetter the wood, the more water to be evaporated and more energy you lose. Burn wood that has dried for at least six months, and your fire will start more easily and give off more heat.

The Oxygen Supply. The second essential for combustion, oxygen, is controlled by the damper or air inlet of stoves. There are many different types of dampers and air inlets, but all of these have the same function; they allow you to control the flow of air going to the fire. Shut the air controls way down and the fire will burn slowly; open them and the fire burns quickly.

But, the air controls not only determine the rate of burning, they also determine the efficiency of the combustion process, because they also affect the third essential for combustion—temperature.

Turning the damper down (Fig. 7-1A), as is often done for long burns, results in a cool fire. Because of the low temperatures, the gases produced by pyrolysis leave the log unburned, and a potentially valuable amount of heat is lost. Compare this with a fire burning with the damper wide open (Fig. 7-1B). The fire is hot enough to burn the gases, but high temperatures create a draft so strong that heat is lost up the chimney before it can be transferred to your house. It may seem as if there is no way to win, so a compromise is necessary.

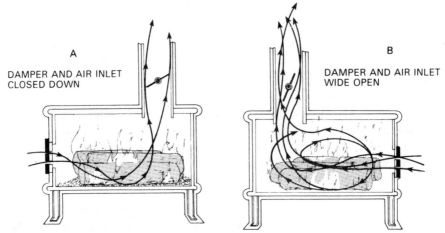

Fig. 7-1: Damper position and how it affects the fire. With wood burning stoves that use a damper, be sure to open the damper before opening the firebox to check the fire or add more wood. This allows accumulated gases to exit up the chimney, possibly preventing them from igniting or even exploding with the sudden rush of air into the firebox.

All we can tell you is that the best position is somewhere in-between these two, and a little experimentation will help you discover where that position is for your particular stove. That is, carefully read the owner's manual that comes with your stove and follow instructions to the letter.

Before Starting Fire

Before starting your first fire in a new stove, read your owner's manual again most carefully for fire burning tips. For example, many stoves require either a two-inch layer of sand or a layer of firebrick to protect the bottom plate of the stove.

Starting a Fire

Nothing can be quite so frustrating to a beginner as failing to start a fire. But once you learn the basic principles, nothing could be easier. Again the three essentials, fuel, oxygen, and temperature, come into play. To start a fire you need dry fuel, plenty of oxygen, and high temperatures.

1. Open the damper and air inlets on the stove all the way.

2. Crumple or shred newspapers and place them near the front of the stove. Burning these will provide the high temperatures needed to ignite the wood.

3. Now stack small pieces (up to 1 inch in diameter) of very dry kindling on the paper. Softwoods like pine and cedar split into thin pieces make the best kindling, but any wood will do if it is dry and finely split. If you use your stove intermittently, it will be wise to secure a supply of pine scraps as kindling.

4. Light the paper, wait until the fire is blazing, and then place full length logs on top (Fig. 7-2A). Stack the logs so that air can circulate freely around them. Do not leave too much space between the logs, as this will interfere with the proper heat reflection between them. Larger pieces of wood and logs often are incapable of sustaining their own flames, but if the logs are properly spaced, they will retain the heat and intensify it to the point of a hot fire. This is why you should avoid one-log fires; two logs are better, and, if your stove will permit, three are best.

5. Allow for ample draft in the beginning. After 10 minutes or so, close down the air controls to the desired location. The fire will now spread slowly backward toward the rear of the stove (Fig. 7-2B).

6. With most airtight stoves, it is not necessary to reload until the wood in the firebox is completely burned up and there are only embers left. Rake the embers up to the front (Fig. 7-2C), and reload as shown in Fig. 7-2D.

Caution: Never use a petroleum product such as gasoline or lighter fluid to start a fire in a stove. This is such a hazardous method of starting a fire that an explosion or flashback may occur and your whole house may become part of the blaze.

Frequently, the cold stove surfaces can have the same effect on a young fire as cold water. Such surfaces, in a phenomenon known as "quenching," cool the flame gases, sometimes to such an extent that the fire is extinguished. One way to overcome the quenching effect is to begin your fire with easy-to-burn materials—crumpled newspaper and very dry, small pieces of kindling. Even in a cold stove, these will burn long enough to warm the stove surfaces to the point where they will not interfere with the burning of larger pieces of wood—those pieces that actually generate heat to the stove and the room. (Make sure all wood is at room temperature.) Another method of avoiding the quenching effect is to bank your fire from the night before by spreading the ashes over the remaining embers. Banking will produce no room heat, but will prevent the stove from cooling to the quenching point.

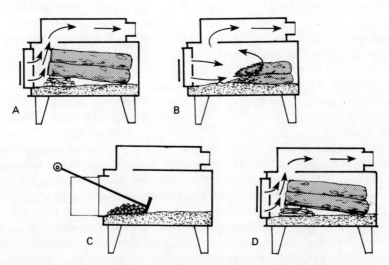

Fig. 7-2: Steps necessary to build a fire in a wood burning stove.

Making the Fire Last

No one likes to wake up to a cold house, so learning how to make a fire last is essential. Here are several tips that will help you accomplish it.

Furnace assist. Set your furnace thermostat at 55 to 60 degrees so that the furnace can take over if the fire goes out.

Thermal momentum. Overheat your room to perhaps 75 degrees F. Then, even if the house cools 15 to 20 degrees by morning, the temperature will still be a comfortable 60 degrees F.

A use for magazines. Glossy magazine paper contains a lot of filler which does not burn well and leaves behind a large amount of ash. Lay 10 to 15 sheets of such paper on the fire. It will burn, leaving a thick layer of ash on top of the wood. This reduces the amount of air that can get to the fuel and slows down the burning rate.

Knotty wood. When you are splitting wood, you will come across large knotty pieces that defy hammer and wedge. So be it, these are overnighters. Their size will ensure they burn slow. Save these and your densest woods for the coldest nights.

Damper shutdown. Close the dampers for long burns. Some of the heat loss due to unburned gases will be compensated for by better heat transfer between the fire and stove walls. Remember that these unburned gases end up as creosote in the stove pipe or chimney. Periodically check your flue or pipe and clean as necessary.

Set a burning schedule. Try to begin a new burning cycle at beginnings of long periods of inattention. Before bedtime, all that should remain of the previous fire is a bed of coals. This ensures that a fuel change will last for the maximum period of time.

Mix green wood. Mixing green wood with dry will extend the wood pile and make the fire last longer.

Sleepwalk. Obviously, the alternative to making a fire last is to get up earlier in the morning to stoke the stove, then go back to sleep.

Ash disposal. One of the byproducts of burning wood is ash, and eventually you will have to remove some or it will fill your stove. Remove ash with a shovel and place it in a heavy metal bucket. Assume that there are still hot coals mixed in with the ashes (there usually are), and do not place them in the garbage in or near anything else combustible. As mentioned in Chapter 5, you may wish to save the ashes in a metal drum; they are an excellent fertilizer, they can be used to give traction on icy walks, and they can even be used to make soap.

There are three other items (Fig. 7-3) considered useful for the safe operation of wood stoves:

1. *Metal bucket.* As just mentioned, ashes are one of the final products of combustion, and these will need to be emptied from your stove from time to time. Concealed within these ashes will almost certainly be some hot coals, ready to spring back to life should they come near anything combustible. Empty ashes only into a metal bucket, preferably with a lid, never into the garbage.

2. *Smoke detectors.* These modern devices are able to detect a house fire while it is still in the smoldering stage. Their track record for saving lives is impressive, and their low cost makes them even more attractive. Install one or two if you have not done so already.

Operating Wood Stoves and Fireplaces   149

Fig. 7-3: Three safety items that should be near every wood stove.

3. *Fire extinguishers.* Large fires begin as small fires. Obviously, you should be prepared to end any fire while it is still in the small stage. A small fire extinguisher, mounted in plain view near an exit that would not be blocked by a fire, could one day save you many times its low cost.

Smoke Conditions

There is an old saying, "Where there's smoke, there's fire." True enough, but just as true is the statement, "Where there's smoke, there's a poor fire." Smoke is the result of the incomplete combustion of the gases, tars, and carbon particles in the wood. Incomplete combustion may be due to green wood, resinous wood (softwoods), a poor kindling effort, a cold stove, or a combination of these factors. Lightweight hardwoods such as aspen (poplar), cottonwood, and willow also tend to produce smoky fires (see pages 17 and 19). Dark, sooty smoke is indicative of a low temperature fire (150 to 500 degrees F). Not only is such a fire inefficient and a waste of wood, but these low temperatures promote condensation of water which mixes with the soot and tars to form creosote, a tar-like substance which may result in dangerous stovepipe fires.

In addition to making certain that you are using the proper materials, and in the proper condition, another way to reduce or eliminate smoke is to buy a stove with a secondary air inlet (Fig. 7-4). A secondary air inlet admits extra air at the point where incomplete combustion is taking place—in the flame area. The extra air facilitates more complete combustion of the gases, tars, resins, and carbon particles, reducing soot buildup and lessening the danger of a creosote fire.

Preventing Creosote Buildup

As stated in Chapter 5, one of the problems of wood combustion is the formation of creosote. There is no one woodburning practice, nor is there just one type or condition of wood which promotes creosote formation. All woods contribute to creosote formation, although some are more guilty than others, and no matter how careful you are in controlling your fire, you will get some creosote buildup. Below are listed the major causes of creosote formation (Fig. 7-5):

1. *Burning green wood,* especially in large amounts, promotes quick creosote buildup because the excessive water condenses on the pipe and flue walls.

2. *Burning softwoods* promotes quick creosote buildup because they are high in resins which are a major component of creosote.

3. *Incomplete combustion* of the gases means that resins and oils are carried

to the pipe and flues where they are deposited, adding to creosote buildup.

4. *Low-temperature fires* (150 to 500 degrees F) promote creosote buildup. Below 250 degrees F, creosote forms rapidly; and below 150 degrees F, moisture condenses, adding to creosote buildup.

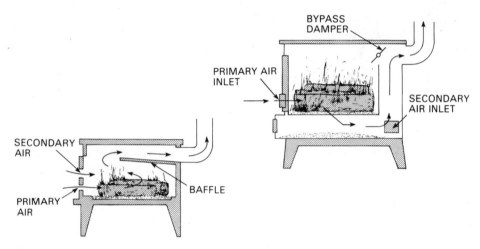

Fig. 7-4: Typical stove with a secondary air inlet. The secondary air strikes the flame, resulting in more complete combustion of the gases and other materials.

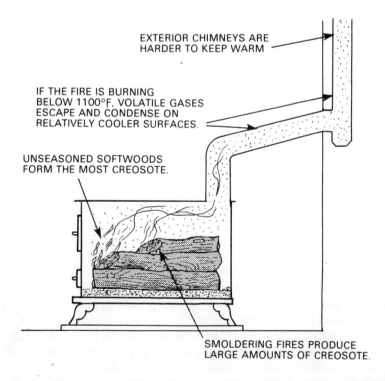

Fig. 7-5: The three common causes of creosote formation are generally incomplete combustion, cool surfaces, and wet wood.

5. *Exterior (cool) flue surfaces* promote creosote buildup. Build all flues inside the home, and insulate those built outside the home.

6. *Late stage fires,* marked by embers or red coals, mean a hot fire in the stove or fireplace but cool pipes and flues, thus allowing for creosote formation. Dense smoke carries a high amount of creosote-forming ingredients.

Obviously, any practice that is opposite of those which contribute to creosote buildup will tend to reduce creosote formation. Burning seasoned wood, hardwoods, maintaining a hot fire (frequent stoking), interior pipes and flues, all retard the formation of creosote. Ironically, as mentioned in Chapter 5, the modern, efficient airtight stoves actually promote creosote formation. These stoves, with their baffles and secondary air, more completely utilize the combustibles in the firebox and radiate more heat to the room. This reduces the amount of heat entering the pipes and flues, increasing the chances of creosote formation. If you own one of these high-efficiency units, one method suggested by most experts to reduce creosote buildup is to deliberately have a hot fire for 15 to 30 minutes each day. This hot fire tends to burn off the creosote in very small amounts each day, thus eliminating the buildup problems. (Never deliberately start a chimney fire to remove creosote from a system.) Other methods of getting rid of creosote deposits are given in Chapter 8.

Dangers of Creosote Buildup. The most serious danger of creosote is a flue or chimney fire. Creosote in small amounts is not dangerous, but large amounts of creosote may suddenly burst into flame when the pipe or flue gets very hot. A chimney fire may cause burning debris to shoot onto the roof (even if you have a screen), the stovepipe may glow a crimson red, and flues may burn, crack, or warp. In some instances, dangerous vibrations may result from a chimney/flue fire where all of the local oxygen is instantly consumed in an explosive convulsion, the fire goes out briefly, only to be reignited when the oxygen is replenished. This on/off cycle produces a violent vibration in the stovepipe, possibly causing it to break loose, showering hot debris all over the house, and anybody who happens to be nearby. In the flue, the on/off cycle could not only crack the flue, but could damage the masonry as well.

The best way to control a chimney fire is to cut off the air going to it. If you have made sure your chimney is sound and airtight, if the stovepipe and stove connections are tight and secure, and if there is only one stove connected to the flue, you can control the air by shutting the air inlet or the solid flue damper. Obviously, the importance of proper stove installation becomes paramount. If a chimney fire should occur, follow these steps:

1. Call the fire department *immediately.*

2. Close all openings and draft controls if you have an airtight stove. Close the stove pipe damper in a non-airtight stove.

3. If the fire is still burning in the stove, douse it with a rolled-up wet newspaper, baking soda, coarse rock salt, or a squirt from a fire extinguisher. The steam from newspapers or the gases from the rock salt or baking soda will travel up the chimney and will often extinguish the flame. Do not run water into the stove or chimney.

4. Check for flames around the chimney. Wet down the roof and adjacent areas to prevent the fire from spreading.

Frequently, a chimney fire may die out before the firemen arrive. Ask them to check over the chimney and stove to be sure there has been no damage that would make them unsafe to use. In other words, never assume that once the chimney fire has burned out, your problems are over. After a fire, the chimney should be checked carefully before it is used. The high temperatures can crack the liner, and violent forces can shake it apart. Without careful inspection, a second chimney fire may finish the job the first one started.

OPERATING A FIREPLACE

A fireplace is a source of warmth and pleasure if the fire is well built and the right wood is used. To make the task of operating a fireplace easier, there are several accessories (Fig. 7-6) you should have available. For example, it is possible to build a fire directly on the hearth, but it is a great deal easier to use andirons or a grate. These devices raise the fire several inches above the hearth, permitting air to enter beneath the fire and to circulate upward for a faster start and better flames. Tongs (to handle burning logs), a poker (for stirring the fire to help the burn), a shovel (to remove ashes), a broom (to sweep away debris), bellows (to furnish air), and long matches (to light the fire) are other items that will come in handy when tending a fire in your fireplace. Of course, you should have a safety screen or glass doors (see page 115) in front of the fireplace to keep sparks and bits of burning wood from falling beyond the limits of the hearth.

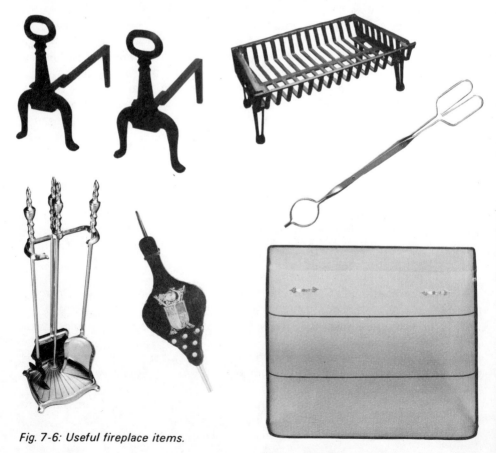

Fig. 7-6: Useful fireplace items.

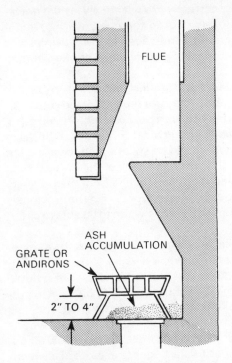

Fig. 7-7: Details of the firebox just prior to building a fire. Leave an ash accumulation of 2 to 4 inches.

To burn wood easily, the proper arrangement of the logs is very important. Place crumpled paper and kindling on the bottom and lay the correct length logs on the andirons or grates. The kindling wood should be exceptionally dry (less than 10 percent water) and very small in size to be of any use. This is because it must be able to catch fire from a match, which obviously does not produce much heat. Chips, twigs, sticks, pieces of bark, old wooden shingles, lumber scraps, etc., make fine kindling. You can never have too much kindling. **Remember: Keep your kindling dry!** The slightest moisture will ruin your kindling for days. If your logs are reasonably dry (20 percent or less water), they can get a little damp and still burn if you have a hot base from the kindling.

When placing kindling in the firebox, do not remove the old ashes. They serve to insulate the fire from the heat absorbing brick base and also help to deflect the draft from underneath the wood right into the wood where it does the most good (Fig. 7-7). (It makes little difference whether you use a grate or an andiron, the principle is the same.) Do not let the ash level build too high; 4 inches is the maximum.

Do not stint on kindling wood. Beginners, thinking in terms of conservation and waste, throw a few chips or twigs into the firebox and think they have a good base for their logs. They are mistaken; the chips or twigs will burn quickly, but will not produce enough heat to ignite the logs. Remember, your kindling will have to heat your logs to at least 300 to 400 degrees F to start them burning, so you are going to need plenty of kindling. Crosshatch eight to 12 sticks or pieces of kindling. If you are using chips, lay enough to form a complete bed over the ashes or paper (if you are using paper). A good base is the key to an easy start and a good fire.

When using newspaper, it must be used correctly. Do not simply lay flat newspapers in the firebox. The layers, too closely packed together, will not admit oxygen and thus will not burn. Instead, crumple the pages loosely, using six to 10 half sheets. Again, do not stint on newspaper; you will need enough heat to get the wood to the burning stage.

Once you have your crumpled newspapers and kindling materials in place—be sure they are spread evenly over the firebox—you are ready to place your larger logs (Fig. 7-8). Do not attempt to start the fire with only one log—it will burn, but produce little heat for your room or cabin. Use at least three logs. A back log from 8 to 10 inches in diameter (a larger log should be split to produce a section this size) should be placed about 1 to 2 inches from the back of the firebox. Next set a fore log, about 4 to 6 inches in diameter, about 3 inches from the back log. Then, set another log about the same size as the fore log between the previously set logs. Avoid squashing the logs together; you want oxygen to be able to circulate about the fire materials.

Once you have all your materials in place and have made sure that the damper is open, strike a match and quickly set it to the crumpled paper at the right, center, and left. If you have a good draft, the paper will quickly catch on fire. Should your attempts at fire building result in smoky backups, it would be wise to preheat the flue. To do this, loosely crumple a full newspaper sheet, light it, and hold it with the hot end at the flue opening, as shown in Fig. 7-9. (Be sure to wear a heavy glove on the hand holding the paper.) The rising heat will clear cold air out of the flue. Use the remainder of your paper to start your fire. **Note:** This technique is not required for all fireplace fires, but is advisable for older fireplaces, for poorly constructed fireplaces, or for slow-drawing chimneys.

When the fire first starts to burn, do not stand too close to it. Also, do not attempt to add more materials to it either. Never leave a beginning fire unattended—sparks, bits of paper, or kindling could pop out over the hearth area. Place a screen in front of the fire, especially during the flaming or "cracking" stages, as this is the most likely time for a flying spark to start a fire in the house. **Caution:** As with fires in a stove, never use flammables such as gasoline, kerosene, or naphtha to start a fire in your stove. In a confined place, the use of such liquids could result in a general conflagration, injury, or death.

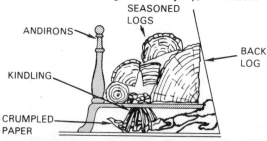

Fig. 7-8: Three or four logs make a good fire. Place a large log on the back of the grate to radiate heat into the room and provide heat to keep the other logs burning brightly. Then put a log in front of the "backlog" and lay kindling—twists of newspaper, dry twigs, or dry scrap lumber—in between. Add the other logs over and around the kindling. When adding a log, rake the coals toward the front of the grate and add new logs to the rear. Encourage a reluctant fire by adding more kindling and adjusting the position of the logs with the poker to get a good draft.

Operating Wood Stoves and Fireplaces   155

Fig. 7-9: Before lighting a fire, always be sure the damper is open. Then, light a newspaper and hold it up toward the flue to counter down-drafts and help prevent smoke from entering the room. Avoid burning the colored pages of newspapers or magazines, as they can release dangerous amounts of lead.

Three-log fires, if they are well made with hardwood logs, can last for a fairly long period. The three logs feed on each other and concentrate the heat. At the point where the logs are red-hot halfway through, place another log on the fire. Always place it at the back, and never put a large log on top of a small one. **Caution:** Wear gloves during this operation (Fig. 7-10). Do not drop the logs on the fire; they will scatter the developing embers, and some of these embers could hit you, causing severe burns.

While it is definitely advisable to start a fire using the driest wood you have, it is neither wise nor even desirable to ignore your green wood (wood with a water content between 20 and 40 percent). **Note:** Never attempt to use freshly cut wood or wood with a water content above 40 percent; the use of such wood will do far more harm than good. After the fire is well underway, you might try placing a green log over the bed of developing coals. The embers will drive off the water and allow the wood to reach a temperature enabling it to burn. Be sure your fire screen is in place while burning green wood; the high sap and moisture content trapped inside the wood cells is heated, expanded, and finally exploded, sending sparks all over.

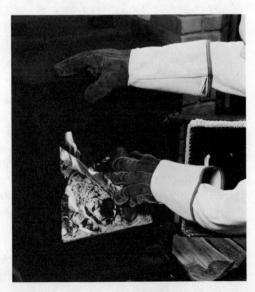

Fig. 7-10: Fire-resistant gloves should be worn when loading a fire.

If there is unburned wood remaining at the end of an evening burning session, it is a good idea to stack it on end in the back corners of the fireplace before you go to bed. In an on-end position (Fig. 7-11), the logs will soon go out and you will have some good dry wood for the next fire. The glowing embers should be covered with ashes, the damper should be left in an open position, and the fire screen should be closed. Use fire tongs when stacking the wood, since there are always some glowing embers. Also, remember to close the damper the next morning to prevent any room heat from going up the chimney.

To keep a fire throughout the night, turn the remaining log(s) with a poker or tongs so that the side with the embers is toward you, and then fully cover this log with the ashes. Ashes cannot burn at temperatures in the fireplace, so they will form a protective blanket over the embers. In the morning, simply rake away the ashes, and you have the embers to start a new fire—use dry twigs and small logs (1 to 3 inches). Meanwhile, the encased embers will enable the fireplace bricks or metal to throw heat during the night.

Fig. 7-11: Setting logs on-end at the end of a burning session.

Other Methods of Starting Fireplace Fires. While the kindling method of starting a fire requires the greatest amount of "know-how," it is ultimately the most satisfactory. There are, of course, other means of starting fireplace fires. For example, there are electric starters. These are similar to the electric charcoal lighter, but are somewhat heavier duty. Your fireplace store will be able to provide you with more details. Follow all manufacturer's instructions concerning installation, use, and maintenance.

While *small* amounts of charcoal can be used to start a fire, it is best to **never** use charcoal in your fireplace. Most processed charcoal products, such as briquettes, contain chemical substances that could prove toxic under enclosed

Fig. 7-12: One method of applying a commercial colored flame maker. Make sure the fire has a good draft before adding the chemical mix.

room conditions. Many charcoal producers warn not to use their product in confined areas.

There are several processed logs on the market. These logs are usually made of sawdust, wax, and a high-temperature binder. (Some even contain chemical coloring agents that give them a colorful effect when burning.) Processed logs usually burn without excessive flames for up to three hours (depending upon draft conditions) and because of the nature of their binders, will hold together throughout the entire burning period. Most of these logs are easy to ignite; use several crumpled sheets of newspaper under the logs and light. As a general rule, most manufacturers do not recommend adding their logs to glowing coals or a flaming wood fire. Also they should be used only in brick, metal, or ceramic fireplaces; they are not designed for use in wood-burning stoves or barbecue equipment. Furthermore, do not cook over a fire of these logs.

There is equipment available in which you can make your own synthetic logs from newspaper. These devices roll newspaper into logs of up to four inches in diameter. Burning time of newspaper logs varies depending on their size, how tightly they are rolled, and on draft conditions.

There are synthetic logs and similar items on the market. Check with your fireplace store for such products. But, when using any of them, carefully read and follow all manufacturer's instructions.

Adding Color to Your Fireplace Fire

The yellow-orange flames of a wood fire can produce an almost hypnotic effect as they curl through, around, underneath, and over the burning logs. Ordinarily, these flames are adequate to give pleasure to most people most of the time. There may be certain occasions—parties, year-end holidays, etc.—where fireplace owners may wish to create special effects by means of colored flames. The entire affair will be enlivened, and everyone's enjoyment will be enhanced by viewing your fireplace fire displaying a miniature "fireplace rainbow."

The yellow-colored flames in ordinary wood are due to the sodium chloride (table salt) and calcium chloride in the wood, plus partially-burned volatile material. For multi-colored fires, the problem is to find materials that will produce the colored effects. Such materials are not common. Saltwater driftwood will produce a blue-lavender flame, and applewood, aged four to five years, will produce multi-colored flames. But these woods are scarcely in sufficient supply to satisfy all those who wish to have colored effects in their fireplaces. There are "chemicals" available at your fireplace and stove supply centers that will produce various colored flames (Fig. 7-12). Use them as directed by the manufacturer.

Wood Burning Appliance Maintenance and Safety 8

For safe and efficient operation, any wood burning appliance must be properly maintained. The stove or fireplace and its chimney must be kept clean. The creosote must be kept in check to reduce the possibility of chimney fires to a minimum. Any wood burning appliance requires a little work, but its reward is plenty of safe and efficient heat. A good maintenance program also eliminates most of your operating problems.

SMOKY FIRES

One of the most common problems of wood stove or fireplace installations is smoky fires. Smoke may come into the room through the openings in the stove or the fire may not burn properly because it lacks an adequate draft. Six main causes and cures are:

1. *Wet wood.* Green or wet firewood causes smoke problems as much of the heat of the fire is used to dry the wood. The cure is to keep a hot fire going and to use seasoned dry wood. If green or wet wood must be burned, split it finer and mix it with dry wood. Softwood may cause smoky fires because of the resin in the wood.

2. *Flue too small.* The stovepipe and chimney flue must be large enough to carry the smoke and gases outside. Follow manufacturer's recommendations for stovepipe size. Do not reduce the pipe size between the stove and chimney connection. An 8 by 8 or 8 by 12 inch chimney flue is usually the minimum size for a chimney. If two wood stoves are connected to one chimney, a larger flue may be needed.

3. *Flue too large.* Many older houses have a large central chimney with several fireplaces and flue openings. If this chimney is used with only one stove or heater, there may not be adequate draft to keep the column of smoke rising. By reducing the cross sectional area of the top of the chimney or installing a stovepipe through the center of the flue, the smoke problem should be solved.

4. *Obstructed flue.* Often stovepipes or flues become partially filled with soot and creosote, especially with small or slow fires. Cure this problem by checking flues and stovepipes once a month during the heating season, and clean them when a buildup starts to occur.

5. *Downdrafts.* Nearby trees, buildings, or roof projections often cause downdrafts during windy periods. Raising the height of the chimney, removing the obstruction, or placing a cap on the chimney may correct the problem.

6. *Lack of oxygen.* A fire needs oxygen to burn properly. In a tight, well insulated house, infiltration has been reduced to a minimum. This lack of air can sometimes cause smoke to be pulled back into the house through an adjacent

flue. Opening the basement door or a nearby window an inch or installing an air intake to the stove area will generally eliminate this problem.

CHIMNEY MAINTENANCE

The chimney can be a source of wood burning problems unless it is properly maintained.

Chimney Inspection

Since the formation of creosote is an unavoidable by-product of burning wood, cleaning chimneys is also an unavoidable chore. It is generally agreed that frequent cleaning is best, but how frequent depends on your circumstances. Check your flue every couple of weeks when you first use it, to establish a rate of creosote formation.

As the rate of formation becomes better understood, it may be possible to lengthen the interval between inspections. If the conditions of operation change, such as delivery of a new batch of wood, return to the two-week inspection schedule. Many installations require cleaning monthly, but in no event should you wait more than one year between cleanings.

The chimney may be inspected from the roof or, in some cases, a mirror can be used up through the chimney flue (Fig. 8-1). If the buildup of creosote and soot is 1/2 inch or more, the chimney should be cleaned. Keep in mind that the greatest accumulation of creosote occurs in the last few feet—the part exposed

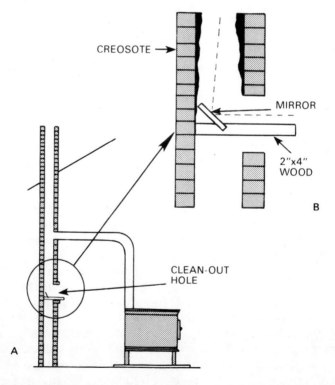

Fig. 8-1: These methods are effective for checking a chimney if it is straight. As shown in B, a scrap piece of 2 by 4 and a small mirror can be combined to check for obstructions in your furnace or fireplace flue. Saw a narrow slot in the 2 by 4 and insert the mirror.

Wood Burning Appliance Maintenance and Safety 161

to the outside temperatures above the roof. As stated earlier, creosote condensation is naturally heavier where the stack is cold. If there is a small deposit of creosote at the point where the stovepipe enters the chimney, it is safe to assume that heavier amounts will be farther up the chimney where flue gas condensation is greater. Ironically, the more efficient your stove, the more frequently your flue will need cleaning.

While inspecting the chimney, look for symptoms of deterioration. Loose or crumbling mortar must be removed or replaced. Cracked brick must be removed and replaced. If these conditions are widespread, your chimney may have to be extensively repaired or completely rebuilt. Cracks in the flue liner must be repaired, but unless you are a skilled mason, leave these internal repairs to a professional. A new chimney liner for a stove, of course, can be installed as described in Chapter 6.

Chimney Cleaning

As mentioned in Chapter 7, some people clean their chimneys by burning a very hot fire for a period of time each day, and this seems to work. **Never** intentionally create a small chimney fire; this may crack the flue liner.

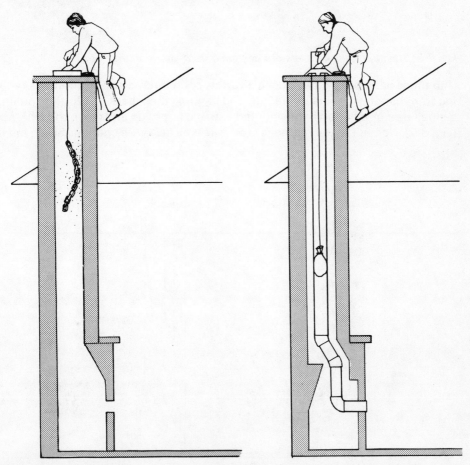

Fig. 8-2: Several older methods of cleaning a chimney.

162 *Firewood and Your Chain Saw*

Chemicals are available at wood burning appliance stores that can be used as a chimney cleaner. These chemicals combine with water released from a hot fire to form a weak acid that dissolves small amounts of creosote. However, do not rely on chemical cleaners as your sole cleaning method. These are best used frequently before creosote becomes too thick.

Eventually every chimney will require manual cleaning. There are commercial chimney sweeps who will do the job for you. Many of these professionals carry large vacuum cleaners to clean up. Many times, however, it is simple enough for you to do-it-yourself. If you clean the chimney, be very careful when climbing on high, steep roofs.

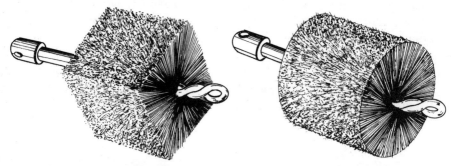

Fig. 8-3: Steel chimney brushes are designed to fit your chimney flue.

In times past, people have used a variety of chimney cleaning methods ranging from hanging chains in the flue and banging them back and forth to pulling a small tree or a burlap bag weighted with old bricks up and down the chimney (Fig. 8-2). None of these methods works as well as a wire chimney brush made

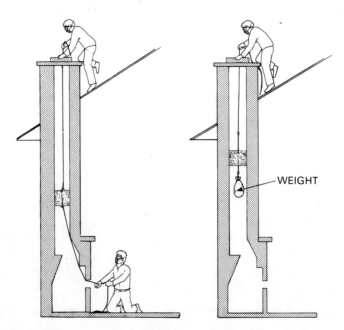

Fig. 8-4: Two methods of cleaning a chimney with brushes.

just for chimney cleaning. These are sized and shaped to exactly fit almost any flue (Fig. 8-3). They can be pulled up and down the chimney by two ropes (one person pulls up from the roof, one person pulls down from below) or by a weight and rope (Fig. 8-4). The easiest way to do the job is with a brush and extension handles. This method allows cleaning from below by one person. To make the job easier, one can install cleanout T's in the stovepipe or chimney. A well placed cleanout T can make chimney cleaning a five minute job (Fig. 8-5).

To clean the chimney, push the brush into the flue. Add a brush extension and continue pushing the brush up the flue. Just a few minutes of alternately lowering and raising the cleaning brush should result in a rewarding pile of black chips of soot and creosote. Continue until the flue looks shiny and smooth.

If the chimney is equipped with a cap, screen, or similar device, be sure to keep it clean. A dirty screen can reduce the draft, making fires difficult to light or causing smoke to backfire into the house. If you do not have a screen, consider getting one. While they do need occasional cleaning, they prevent twigs and other debris from falling into your fireplace, and this prevents birds from nesting inside the chimney—another source of draft problems. If you do decide to get a screen, get a sturdy one with a large mesh (1/2 inch).

You will probably find that a few things working together will keep your chimney clean without much trouble. Remember that it is easier to inspect and maintain your chimney frequently than let it wait for one big job.

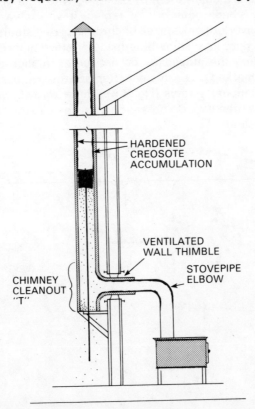

Fig. 8-5: Entry to flue is made easier through a cleanout "T" at the base of the flue.

Stovepipe Maintenance

As already mentioned, stovepipes need periodic checks. One method for checking stovepipes is to tap on the pipe with a metal object. The sound will change from a metal ping to a dull thud as materials build up inside the pipe.

Before disassembling the stovepipe sections, mark them by scratching a number near the end or by taping masking tape on the sections and numbering them. Do not forget to spread a drop cloth or old newspapers on the floor. Also, cover furniture and furnishings, or remove them from the room. The wisest idea is to do your cleaning outdoors. Cover the top of the flue to avoid losing heat from the house. Even if it is summer, unexpected gusts of wind could send loose soot and ash all over the interior of the house. Also, keep in mind that stovepipe ends are sharp, and you can be certain that there is plenty of soot around, so be sure to wear work gloves and a face mask.

Disconnect the most suspect section. If the sections do not come apart easily, do not use excessive force or pound with a hammer; you could dent, bend, or put a hole in the pipe. Two people pulling in opposite directions may help. Loosen the caked-on creosote and ash by working the joints back and forth or by twisting in a left-right pattern until the sections come apart. There are many solvents on the market which loosen tars, creosote, and caked ash. Apply the solvent liberally and let it set at least 20 minutes so that it can do its work. Once you have the sections apart, clean them by brushing with your wire brush. If any section of stovepipe is badly deteriorated, replace it with at least 24-gauge steel (see page 126). Avoid bright galvanized or chrome-plated steel stovepipes since they do not radiate heat as well as blue-oxide or stove blackened pipe.

When reassembling the stovepipe, do not forget to align according to the numbers you have marked on each piece. Replace the damper first, joining the sections with self-tapping screws (Fig. 8-6). Three screws will hold the pipe much more securely than two. Do not reattach the pipe to the stack or flue until that part has been cleaned.

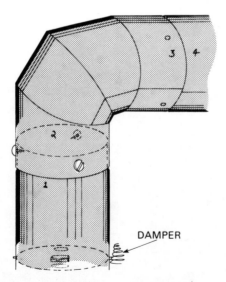

Fig. 8-6: Replacing the stovepipe with self-tapping screws.

Cleaning the Stove

Before commencing any cleaning or repairing of your stove, consult your owner's manual. It will contain valuable information concerning the various parts, where they are located, and how to take them apart. It will tell you of special problems you may encounter and may contain warnings which you should heed, not only to clean the stove, but also to avoid damaging it and possibly to preserve the validity of your warranty.

Never work on a hot or even a warm stove. If the stove is warm, do such preliminary things as laying the drop and assembling the tools. Then, put on your face mask and gloves. Remove the ashes carefully as there may be some hot embers, even if the stove is cool. Also, if there is sand at the base of the stove as instructed by the manufacturer, it is usually a good idea to leave a few ashes rather than disturb the sand. If you do remove some sand, replace it with sand designated to be used in stoves. Do not use sand designed for use on roads, as it may contain salt which is destructive on iron or steel.

Using your trouble light, carefully examine the interior of the stove to see what needs to be done. Place several layers of old newspaper over the base to catch the soot and other material you scrape off. Then, scrape or brush the entire area thoroughly. A shop vacuum cleaner makes soot pickup easy (Fig. 8-7). Dispose of these scrapings. Visually check cracks or corroded areas; mark such spots with chalk. Even if you cannot see cracks, they may be there. Try the "flashlight test." After dark, place a lighted flashlight in the firebox and close the stove door. Turn out all the lights in the room. If you can see any light where the stove is supposed to be solid, you know you have a crack. Mark it with chalk so that repairs can be made. **Note:** Check your warranty card to see if the manufacturer guarantees that part and whether the time limit has run out. Even limited warranties may have a lifespan up to 10 years.

One advantage of doing your stove checkup in spring is that you have plenty of time to order parts or make repairs. Many replacement parts may take weeks, even months to arrive.

After checking the firebox for cracks, check the firebricks (if your stove is so equipped) for damage. Cracked bricks can be repaired with a cement specifically prepared for firebricks. Replacement bricks may be purchased from your dealer or building supply stores.

Sheet Metal Stoves. If your stove is made of metal plates, survey it for warping, dents, or general deterioration. Seriously warped plates, or ones that show definite signs of deterioration, must be replaced. Minor dents and moderate warping are not really serious; they are mostly a cosmetic detraction. Serious dents, on the other hand, should be tended to. Use a rubber mallet to pound out dents; never use a metal hammer.

Occasionally, cracks will develop due to thermal (heat) expansion. Such cracks, as well as all separated joints, should be sealed with stove or refractory cement. If access to a machine shop is available, you might see about having the plates welded. If you are uncertain as to the condition of your stove's firebox, have your dealer or a professional check it out.

Baffles. Many of the modern airtight stoves are built with baffles, which are great for retaining heat in the stove, but are a pure headache when it comes to cleaning. There is usually only a small space between the baffle and the firebox top (Fig. 8-8). A long-handled brush or scraper will be needed to clean the baffle. (If you do not have one, tie your scraper or brush to a stick or a section of an old broom handle.) Some baffle-style stoves are equipped with push-out panels which may be removed to facilitate cleaning. If the panels are hard to remove, they may be cemented. A sharp tap with the hammer (plus a block of wood) may remove it. Some baffles are bolted or screwed on, and these may be removed to make cleaning easier. A little household oil will make future removals easier. Scrape off all old cement before applying new cement and reassembling the baffle to the stove.

Draft Controls. Generally, draft controls do not give the stove owner too many problems. However, these controls may become caked with ash, soot, or just plain dirt. Chemical solvents can be used to soak away caked materials. If the lever or handle on your draft control tends to get as hot as the fire itself, you might consider fitting it out with a lever that does not conduct heat.

Fig. 8-7: Removing soot from a stove with a vacuum cleaner. Remember that it is far easier to remove soot (and even creosote) immediately after a heating season than it is to wait until fall when the materials have had a chance to solidify for several months.

Doors, Gaskets, and Seals. Doors, and their component parts, get lots of use (and some abuse). Check for loose or broken door hinges or latches. Even if the door still closes, a loose hinge or latch will cause the stove to lose its airtight qualities. Even if your stove is not of the airtight variety, a loose door will eventually come completely off, and usually at the worst possible moment—when the temperature outside drops. A neglected handle could result in you holding a separated handle with a hot door waiting to be opened. When replacing, get a handle with independent coils (Fig. 8-9) which do not get hot when the stove is hot. A worn gasket will play havoc with an airtight stove's efficiency. If a gasket is frayed or worn, *do not* attempt to sand or file it; replace the gasket with a new one, and dispose of the old one immediately.

Thermostats. If the stove is so equipped, the thermostat is needed to regulate the fire to produce a long-lasting, steady heat. But, many stove owners neglect the thermostat until it no longer works, and then it is usually too late to do any-

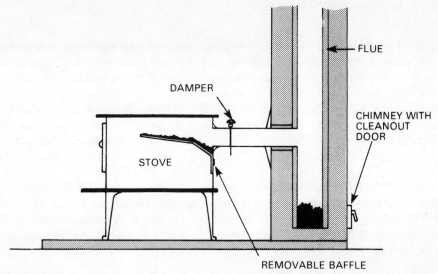

Fig. 8-8: Do not neglect to clean the top of the baffle. Many stoves have a removable baffle.

thing. As described earlier, thermostats are delicate mechanisms made of two strips of different metals which expand at different rates, alternately opening and closing the damper as the temperature rises and falls. To check your thermostat, move the control lever to the closed position and see whether the flap closes securely over the air intake passage. Uneven operation of this component while the stove is working is a tell-tale sign of deterioration. The metallic strips may be pushed together temporarily, but when your thermostat falters consistently, it must be replaced.

Dampers. Dampers regulate the draft (air flow) of the stove system. If they become clogged with ash, soot, or creosote, they cannot function and you cannot expect a decent fire. Regardless of the type of damper the stove has, it needs careful attention. Remove all polluting materials. Be careful when working

Fig. 8-9: Stove door handle with independent coil. Handle remains cool even though stove may be hot.

around dampers as they are delicately balanced, and heavy-handed cleaning may actually do more damage than good. Here is a place where solvents will come in handy. After the damper has been cleaned, a few drops of household oil will insure a season of smooth operation.

Cookstoves. When cleaning a wood burning cookstove, sweep or vacuum the area under the cooking plates and empty the clean-out chamber, which is usually found under the oven (Fig. 8-10). This is generally reached by lifting out the nameplate generally located below the oven door or from a cleanout door on the side.

Floor and Wall Protectors. Too many wood stove owners neglect their floor and wall protectors during the process of inspection and repair. This is a serious mistake. During the course of time, your protectors may have worked loose from their moorings. Check all fasteners. Check for corrosion, warping, cracked and loose mortar and brick.

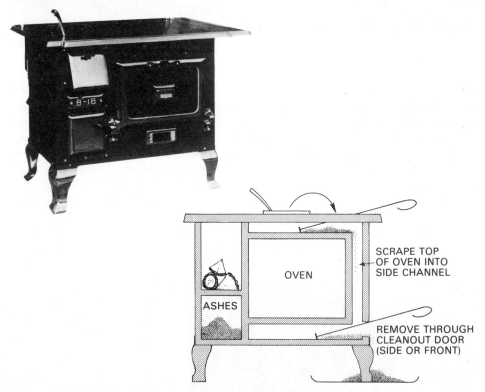

Fig. 8-10: Method of cleaning cookstoves.

The Legs. The legs hold up the stove. If anything should go wrong with them, the stove could topple over, scattering hot coals all over the floor with disastrous results. If any of the legs are cracked, replace them or have them welded by a professional. Check for loose or missing bolts; tighten or replace. Also, check the floor mat to see if the legs may have worked their way through. A worn spot is a weak link in your overall protection—a place where a spark or cinder could penetrate, setting off a fire.

Fig. 8-11: Applying stove black to a stove.

Summer Storage

If your stove is not to be used over the summer, remove all ash and sand to prevent corrosion. If any rust spots or shiny places have developed, these must be treated to prevent further deterioration. Remove the rust with emery cloth and wipe away the dust. Stove black will not only inhibit further rust formation but will make the stove look like new. Stove black also makes the stove more efficient as black radiates heat better than light colors (Fig. 8-11). A coat of cooking oil over the cooking surfaces will protect these areas even more. **Note:** Do not let the oil spill onto painted surfaces, making it impossible to paint these areas. An open box of baking soda will absorb moisture, retard rust and mold formation, and keep the stove sweet smelling (Fig. 8-12).

Fig. 8-12: Baking soda will absorb moisture and retard rust and mold formation.

If the stove is to remain in place, cover it with heavy plastic to keep out dust and moisture. If you move the stove, do so carefully, especially if it is cast iron, as dropping or hitting hard objects against it may crack the plates. Do not store the stove in the basement, as these places are damp. If it must be stored there, place a dehumidifier in the room. If the stove is to be placed on the porch, cover it with heavy plastic to keep out dust and rain. A storage shed is the best place for the stove, but even here it is best to cover the stove with plastic.

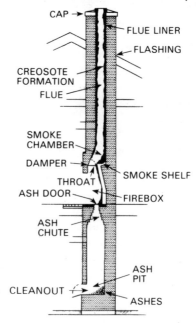

Fig. 8-13: Diagram of complete fireplace-chimney unit.

FIREPLACE MAINTENANCE

The fireplace requires little maintenance. In fact, most of the problems include (Fig. 8-13):
1. Excessive soot and/or creosote buildup in the flue;
2. Blockage, loose bricks, damaged flue liner anywhere in the chimney;
3. Loose mortar or cracked bricks in the firebox;
4. Improperly functioning damper; and
5. Unclean firebox.

The first two maintenance procedures are accomplished in the same manner as for wood burning stoves given earlier in this chapter.

The Firebox. Before attempting to build a fire, check the entire firebox for sources of trouble—loose or cracked bricks and crumbling mortar. Cracked bricks should be replaced. Be sure to use firebricks which are made from special clays and are specially fired to resist extremely high temperatures. Also, ordinary mortar must not be used. Instead, use either a special firebrick mortar or an aluminum oxide mixture made specifically for use in laying firebrick. If you are not experienced at laying brick, have a professional do the job. It may cost more, but you have the security in knowing that your fireplace is in working order.

The Damper. Faulty dampers cause more trouble than almost any other part of the fireplace/chimney system, and there is no excuse for any of it. Dampers are relatively easy to inspect and to keep in good repair. First, check your controls, whether they be handles, chains, or levers, to see that there is full movement. If your control moves just a little bit, the chances are that there is a blockage in the damper area.

After much use, sundry debris such as soot, creosote, loose mortar, or just plain dirt may have accumulated on the smoke shelf behind the damper hinge, preventing the damper from either opening or closing completely. To correct this situation, detach the damper blade from its adjusting control (sometimes just a cotter pin is involved) so you can get behind the damper. Clean the shelf with a wire brush, a stick, or an old whisk broom. Old mortar may present a little problem if it was wet when it fell and has hardened. Remove with a hammer and an iron pipe or a cold chisel. After reassembling the damper, test it a few times to make sure that it is working properly.

Ash Dump. Generally, few problems develop in connection with the ash dump. Old ashes, however, should be cleaned out yearly, even if there is not a big load. Ashes contain organic matter, and foul odors could develop from stale ashes left in a damp environment. When removing ashes, if they are not already damp, wet them down with a bit of water. This prevents powdery particles from scattering into the air and holds down carbon dust. Also, do not throw out the ashes; they are, as stated before, valuable as fertilizer.

Brick. Brick surfaces can become discolored by smoke and soot. To remove this dirt from brick fireplaces and hearths, or stove pads, use one of the following methods:

1. Mix up a cleaning solution of 3 tablespoons of trisodium phosphate to 1 gallon of warm water. Apply it with a stiff brush and scrub hard. Clean the brush often. Rinse the surface well. If the brick is smooth, you can use a sponge instead of a brush. Be sure to wear rubber gloves and safety goggles when using this solution. Also, avoid contact with your skin.

2. For an alternative cleaning solution, add 1/4 cup of ammonia to 1 quart of warm water and detergent. Mix in enough detergent to make a respectable suds. Work on the bricks as described in method 1, and rinse well.

3. Try a powdered cleaner containing bleach. Rinse well.

4. Use a commercial brick cleaner. Follow instructions carefully.

5. If all else fails, scrub bricks with a dry stiff-bristle brush to remove the soiled layer of brick. Take care not to damage the mortar.

Once the brick surface is clean, apply a heat-resistant sealer. This will help prevent further discoloration and will also make the cleaning job easier next time around. Incidentally, the same methods of cleaning just described will work well on stone fireplaces, too.

WOOD BURNING SAFETY

Studies into the causes of fires in homes heated with wood burning units reveal that most could have been avoided. The majority of fires result from improper installation of the wood burner or the chimney. The next most prevalent causes are unsafe operation and poor maintenance. Very few fires have started as a result of a poorly made stove or an unsafe fireplace. **Remember:** The most important factor in burning wood is safety, not saving money or saving energy.

By carefully adhering to the following suggestions, you can prevent the major causes of wood burner fires in homes:

1. Make sure you understand and follow the manufacturer's instructions when installing and operating the wood burning appliances. Check with the manufacturer or dealer who sold you the unit if you are unsure about anything.
2. Check with local code officials about wood burning appliance installation requirements.
3. Make sure there are safe clearances between the stove and combustible floors, walls, and ceiling. Place the stove on a fireproof base or pad.
4. Never locate flammables such as paper, rags, and even fuel wood near a stove or fireplace. Never put logs on top of the stove to dry out. This is a very dangerous procedure.
5. Wooden and upholstered furniture should be kept at least 3 feet from the stove.
6. Never use flammables—kerosene, gasoline, etc.—to start or rebuild a fire. Also, do not use charcoal or other starter fluid. Only use those products designed expressly for wood-appliance lighting.
7. Keep a fire extinguisher in a convenient place near the stove or fireplace at all times.
8. Burn dry, well-seasoned firewood. Never burn poison ivy, plastics, chemically treated poles and railroad ties, or garbage in your wood burning appliance.
9. Always keep a screen in front of a fireplace to prevent sparks from popping out onto floors and carpets.
10. Dispose of ashes in a closed metal container—never in paper bags or cardboard boxes. Even cold ashes may conceal hot coals.
11. Inspect the chimney periodically for creosote buildup and cleanliness. If it needs to be cleaned, do it immediately.
12. Check the condition of the stovepipe occasionally. If the stovepipe is corroding, replace it. If the damper does not operate properly, replace it.
13. Keep a pound or two of baking soda or coarse rock salt stored in a nonmetallic container near the stove; it can be handy in helping to reduce the severity of a chimney fire.
14. Instruct other members of the family on safe and proper operation of the wood burning stove or fireplace.
15. Install smoke detectors to alert family members if a malfunction of the wood burning appliance should occur.
16. Keep small children away from stoves and fireplaces; even when cold, such places are dirty and dangerous for little ones.
17. In the event of a chimney fire, call the fire department immediately and then proceed as described on page 151.
18. Prepare your family for a fire. Plan escape routes from every room in your home, and make your family aware of fire survival techniques. Have a practice drill so that everyone understands exactly what to do in the event of a fire.
19. Keep the fireplace damper open as long as the fire or embers burn.
20. Do not burn household trash in your stove or fireplace. Plastics, especially, can give off toxic fumes.

References

Book of Successful Fireplaces: How to build, decorate, and use them by R. J. Lytle and M. J. Lytle. Structures Publishing Company, Farmington, MI 48024.

Burning Wood. Cooperative Extension, Northeast Regional Agricultural Engineering Service, Cornell University, Ithaca, NY 14853.

Buying & Installing a Wood Stove by Charles Self. Garden Way Publishing Company, Charlotte, VT 05445.

Chimney & Stove Cleaning by Christopher Curtis and Donald Post. Garden Way Publishing Company.

The Complete Book of Heating with Wood by Larry Gay. Garden Way Publishing Company.

Curing Smoky Fireplaces by Douglas Merrillees. Garden Way Publishing Company.

Energy Primer. Portola Institute, 588 Santa Cruz Avenue, Menlo Park, CA 94025.

Fireplaces and Chimneys. Farmers' Bulletin #1889, USDA. Superintendent of Documents, Washington, DC 20402.

The Forgotten Art of Building a Good Fireplace by Vrest Orton. Yankee, Dublin, NH 03444.

Fuelwood and Wood Burning Stoves. The Pennsylvania State University, Cooperative Extension Service, University Park, PA 16802.

Growing Trees for Timber in New York's Small Woodlands by A. Dickson. Information Bulletin 67, Cornell University.

Heating Your Home with Wood by Neil Soderstrom. Harper & Row, New York, NY 10017.

The Lange Stove Catalog and Wood Heat Guide by David Lyle. Scandinavian Stoves, Inc., Box 72, Alstead, NH 03602.

Link a Wood Stove to Your Fireplace by Mary Twitchell. Garden Way Publishing Company.

Modern and Classic Woodburning Stoves by Bob and Carol Ross. Overlook Press c/o Viking Press, New York, NY 10017.

Morso Wood Heat Handbook by Lee Gilchrist. Southport Stoves, 248 Tolland Street, East Hartford, CT 06108.

Using Coal and Wood Stoves Safely. NFPA No. HS-8. National Fire Protection Association, 470 Atlantic Avenue, Boston, MA 02210.

Wood As a Home Fuel. Extension Service, University of Vermont, Burlington, VT 05401.

The Wood Burners Encyclopedia by Jay W. Sheldon and A. W. Shapiro. Vermont Crossroads Press, Waitsfield, VT 05673.

The Wood Burners Handbook by David Havens. Harpswell Press, Brunswick, ME 04011.

Wood Burning Quarterly. Wood Burning Quarterly Magazine, 8009-34th Avenue South, Minneapolis, MN 55420.

Wood Heat by John Vivian. Rodale Press, Emmaus, PA 18049.

Wood Heat Safety by Jay W. Sheton. Garden Way Publishing Company.

Wood Heating Handbook by C. Self. Tab Books, No. 872. Blue Ridge Summit, PA 17214.

The Wood Stove and Fireplace Book by Steve Sherman. Stackpole Books, Cameron & Kelker Streets, Harrisburg, PA 17105.

Wood Stove Buyers Guide by A. A. Barden III. Northeast Carry, 110 Water Street, Hallowell, ME 04347.

Wood Stove Directory. Energy Communications Press, Inc., Manchester, NH 03108.

Wood Stove Know-how by P. Coleman. Garden Way Publishing Company.

Wood Stoves: How to Make and Use Them by Ole Wik. Alaska Northwest Publishing Company, Anchorage, AK 99510.

METRIC CONVERSIONS

inches	milli-meters	feet	centi-meters	feet	meters	feet	kilo-meters
1/16	1.6	1	30.5	25	7.5	5000	1.5
1/8	3.2	2	61.0	30	9.0	5280	1.6
1/4	6.4	3	91.5	35	10.5		
3/8	9.5			40	12.0	miles	kilo-meters
1/2	12.7	feet	meters	45	13.5		
5/8	15.9			50	15.3		
3/4	19.0			60	18.3		
7/8	22.2	4	1.2	75	22.8	1	1.6
1	25.4	5	1.5	100	30.5	2	3.2
2	51.0	6	1.8	200	61.0	3	4.8
3	76.0	7	2.1	300	91.5	4	6.4
		8	2.4	400	122.0	5	8.1
		9	2.7	500	153.0	10	16.2
inches	centi-meters	10	3.0	1000	305.0	20	32.4
		11	3.4	2000	610.0	30	48.6
		12	3.7	3000	915.0	40	64.8
4	10.0	13	4.0	4000	1220.0	50	81.0
5	12.7	14	4.3			100	162.0
6	15.2	15	4.6				
7	17.8	16	4.8				
8	20.3	17	5.2				
9	22.9	18	5.5				
10	25.4	19	5.8				
11	28.0	20	6.0				

Conversion Factors

Measurement and Symbol	Multiply By	To Find Measurement (with Symbol)
inches (in)	25.4	millimeters (mm)
feet (ft)	0.305	meters (m)
yards (yd)	0.9	meters (m)
miles (mi)	1.6	kilometers (km)
sq. yards (yd^2)	0.84	sq. meters (m^2)
acres	0.405	hectares (ha)
sq. miles (mi^2)	1.47	sq. kilometers (km^2)
deg. Fahrenheit (°F)		deg. Celsius (°C)

°F − 32 × .556 = °C

pounds (lb)	**0.454**	**kilograms (kg)**
millimeters (mm)	0.04	inches (in)
meters (m)	3.3	feet (ft)
meters (m)	1.1	yards (yd)
kilometers (km)	0.62	miles (mi)
sq. meters (m^2)	1.19	sq. yards (yd^2)
hectares (ha)	2.47	acres
kilograms (kg)	2.2	pounds (lb)
deg. Celsius		deg. Fahrenheit (°F)

°C × 1.8 + 32 = °F

Note:
1. Use small case letters when writing symbols unless capital letter is specifically called for.
2. No periods are used with symbols. No symbol for "acre."
3. Never make conversion any more accurate than original figure.

Index

Asbestos-cement board, 122, 124
Asbestos millboard, 122
Axes, 43-44, 46

Bucking (felled timber), 68-69
Bureau of Land Management, 56

Chain saws, 27-40, 41-42, 58-69
  chains, 33-34
  construction characteristics, 29-33
  cutting wood with, 61-69
  electric, 39-40
  features, 37-39
  safety, 34-37, 41-42, 58-61, 64-65
    attire, 41-42
    options, 34-37
Charcoal, 157
Chimneys, 129-134, 160-163
  cleaning, 160-163
  inspecting, 132-134, 160
  masonry, 130-132
  prefabricated, 130
Cords (as a unit of measurement), 24, 93
  needed per year, 93
Creosote, 15, 98, 150-152
  preventing, 150-152
Cribs, wood, 74

Felling (timber), 65-67
Fireplaces, 95, 112-114, 116-117, 135-142, 152-157, 170-171
  heat exchangers, 116-117
  installing, 135-142
    prefabricated, 136-142
  maintenance, 170-171
  metal, 112-114
  operating, 152-157
Firewood, 9-20, 22-26, 53-58, 75-92
  buying, 22-26
  choosing, 9-20
  combustion characteristics, 96
  seasoning, 16, 24, 73, 87-88
  sources, 53-58
    commercial, 55-56
    dumps and landfills, 57-58
    public lands, 56-57
    woodlots, 54-55
  splitting, 16, 75-83
  stacking, 83-87
  storing, 88-92
  transporting, 69-72
Fruitwoods, 11, 12-14, 16-20, 25
  burning characteristics, 12-14, 16-20
Furnaces, wood, 95, 109-111
  multi-fuel, 111

Hardwoods, 10, 11, 12-20
  burning characteristics, 12-19
Hatchets, 43
Heat exchangers, 116-117
Heating values, 11-15, 16, 74

Insurance, fire, 119

Kickback, chain saw, 64, 68
Kitchen stoves, 108-109

Limbing, 67-68

Mauls, 47-48
Moisture content (of wood), 11, 15-16, 21, 145

Sawbucks, 44-46, 62
Seasoning (firewood), 16, 24, 73, 87-88
Smoke, eliminating, 149, 159-160
Softwoods, 10, 11, 12-14, 16-20
  burning characteristics, 12-14, 16-20

Solar log driers, 87
Splitters, cone, 50-52
Splitters, power, 49-50, 80-83
Splitting, 16, 75-83
  hand, 75-80
  power, 80-83
Stacking (firewood), 83-87
Storing (firewood), 88-92
Stovepipe, 126-129, 142-144, 164-169
  dampers, 128-129
  installation, 126-128, 142-144
    in fireplace chimneys, 142-144
  maintenance, 164-169
Stoves, see Wood stoves

Trees, 9-11, 12-14, 18-19, 21-22
  anatomy, 9-11
  identification, 21-22

Wedges, 48-49, 67, 76-77
  felling, 67
Wood blocks, 52
Woodlots (thinning), 54
Wood stoves, 95, 101-109, 119-126, 134-135, 142, 145-152, 165, 171-172
  airtight, 95, 102
  box, 95
  cleaning, 165
  design features, 106-108
  draft control, 102-104
  heat capacity, 104-106
  high-efficiency, 95
  installing, 119-126, 134-135
    checklist, 134-135
    location and clearance, 120-126
  installing in fireplaces, 142
  kitchen, 108-109
  materials, 101-102
  operating, 145-152
    starting a fire, 146-147
  safety, 171-172